DISCARD

Evolution and Religion in American Education

Cultural Studies of Science Education

Volume 4

Series Editors

KENNETH TOBIN, *City University of New York, USA*
CATHERINE MILNE, *New York University, USA*

The series is unique in focusing on the publication of scholarly works that employ social and cultural perspectives as foundations for research and other scholarly activities in the three fields implied in its title: science education, education, and social studies of science.

The aim of the series is to establish bridges to related fields, such as those concerned with the social studies of science, public understanding of science, science/technology and human values, or science and literacy. *Cultural Studies of Science Education*, the book series explicitly aims at establishing such bridges and at building new communities at the interface of currently distinct discourses. In this way, the current almost exclusive focus on science education on school learning would be expanded becoming instead a focus on science education as a cultural, cross-age, cross-class, and cross-disciplinary phenomenon.

The book series is conceived as a parallel to the journal *Cultural Studies of Science Education*, opening up avenues for publishing works that do not fit into the limited amount of space and topics that can be covered within the same text.

For further volumes:
http://www.springer.com/series/8286

David E. Long

Evolution and Religion in American Education

An Ethnography

 Springer

David E. Long
Department of Middle, Secondary, Reading
and Deaf Education
Valdosta State University
N. Patterson St. 1500
Valdosta, GA 31698
USA
delong@valdosta.edu

ISSN 1879-7229   e-ISSN 1879-7237
ISBN 978-94-007-1807-4   e-ISBN 978-94-007-1808-1
DOI 10.1007/978-94-007-1808-1
Springer Dordrecht Heidelberg London New York

Library of Congress Control Number: 2011935022

Printed on acid-free paper

Springer is part of Springer Science+Business Media (www.springer.com)

Acknowledgments

My great thanks to those listed below. Through many discussions—some tedious, most spirited—each variably influenced my thinking regarding evolution, science, and religion in American education.

Richard Angelo, Phil Berger, Jeff Bieber, Mary Beth Chrostowsky, Steve Clements, Beth Goldstein, Katherine Hoover, Jane McEldowney Jensen, Stuart Jones, Garry Long, Nancy Long, Gavin McDade, Sarah Neusius, Derek Ruez, Chris Stapel, John Taylor, John Thelin, Ken Tobin, Christina Wright, and John Yopp.

Additional thanks to those who read and commented on drafts of this work as it developed:

Senka Henderson, Jane McEldowney Jensen, Ken Tobin, Kim Sale, and Christina Wright.

Contents

List of Figures

List of Tables

Chapter 1
Prologue: Darwin's Apocalypse

O sinner-man, where are you going to run to?
O sinner-man, where are you going to run to?
O sinner-man, where are you going to run to, all on that day?

Run to the Lord: O Lord, won't you hide me?
The Lord said: O sinner-man, you ought to been a-praying

Sinner-man says: Lord, I've been a-praying
The Lord said: O sinner-man, you prayed too late

Run to Satan: O Satan, won't you hide me?
Satan said: O sinner-man, step right in

Sinner Man (excerpts)
American traditional

Main Entry: **evo·lu·tion**[1]

Pronunciation: \e-və-lü-shən, ē-və-\
Function: *noun*
Etymology: Latin *evolution-, evolutio* unrolling, from *evolvere*
Date: 1622

1 : one of a set of prescribed movements
2 a : a process of change in a certain direction : unfolding **b** : the action or an instance of forming and giving something off : emission *c* (*1*) : a process of continuous change from a lower, simpler, or worse to a higher, more complex, or better state : growth (*2*) : a process of gradual and relatively peaceful social, political, and economic advance **d** : something evolved
3 : the process of working out or developing
4 a : the historical development of a biological group (as a race or species) : phylogeny **b** : a theory that the various types of animals and plants have their origin in other preexisting types and that the distinguishable differences are due to modifications in successive generations; *also* : the process described by this theory
5 : the extraction of a mathematical root
6 : a process in which the whole universe is a progression of interrelated phenomena

---

[1] All entries from Merriam-Webster Dictionary, 2009.

D.E. Long, *Evolution and Religion in American Education: An Ethnography*,
Cultural Studies of Science Education 4, DOI 10.1007/978-94-007-1808-1_1,
© Springer Science+Business Media B.V. 2011

Main Entry: **end**

Pronunciation: \end\
Function: *noun*
Etymology: Middle English *ende,* from Old English; akin to Old High German *enti* end, Latin *ante* before, Greek *anti* against
Date: before 12th century

1 a : the part of an area that lies at the boundary *b (1)* **:** a point that marks the extent of something *(2)* **:** the point where something ceases to exist <world without end> **c :** the extreme or last part lengthwise : tip **d :** the terminal unit of something spatial that is marked off by units **e :** a player stationed at the extremity of a line (as in football)
2 a : cessation of a course of action, pursuit, or activity **b :** death, destruction *c (1)* **:** the ultimate state *(2)* **:** result, issue
3 : something incomplete, fragmentary, or undersized : remnant
4 a : an outcome worked toward : purpose <the end of poetry is to be poetry — R. P. Warren> **b :** the object by virtue of or for the sake of which an event takes place
5 a : a share in an undertaking <kept your end up> **b :** a particular operation or aspect of an undertaking or organization <the sales end of the business>
6 : something that is extreme : ultimate—used with *the*
7 : a period of action or turn in any of various sports events (as archery or lawn bowling)

Main Entry: **world**

Pronunciation: \wər(-ə)ld\
Function: *noun*
Etymology: Middle English, from Old English *woruld* human existence, this world, age (akin to Old High German *weralt* age, world); akin to Old English *wer* man, *eald* old—more at virile, old
Date: before 12th century

1 a : the earthly state of human existence **b :** life after death —used with a qualifier <the next world>
2 : the earth with its inhabitants and all things upon it
3 : individual course of life : career
4 : the inhabitants of the earth : the human race
5 a : the concerns of the earth and its affairs as distinguished from heaven and the life to come **b :** secular affairs
6 : the system of created things : universe
7 a : a division or generation of the inhabitants of the earth distinguished by living together at the same place or at the same time <the medieval world> **b :** a distinctive class of persons or their sphere of interest or activity <the academic world> <the digital world>
8 : human society <withdraw from the world>
9 : a part or section of the earth that is a separate independent unit
10 : the sphere or scene of one's life and action <living in your own little world>
11 : an indefinite multitude or a great quantity or distance <makes a world of difference> <a world away>
12 : the whole body of living persons : public <announced their discovery to the world>
13 : kingdom 5 <the animal world>
14 : a celestial body (as a planet)

This is not a book about the history of evolution nor a history of public receptivity toward it. It is not about questioning the validity of evolutionary theory. This is an anthropology of science education in America today. Why, when we examine the stories and experiences of students from the university, their schooling, and their lives, do we continue as a nation to produce, at least in some quarters, such antievolutionary sentiment? What explicit and implicit rationales do students and teachers employ when choosing to engage or avoid education about evolution? When we step back from the idealistic hopes of science educators and policy makers, how is evolution education handled when the peering spot light of research is turned off?

As the foremost venue by which evolution might be explicated, colleges and schools do much to codify what counts as official knowledge. But as is known, evolution education is not presented evenly within the USA, and in many cases is avoided entirely. If this were not so, state curriculum standards would not waver from including the concept (Lerner 2000). Educational standards, curriculum, and the teachers which enact them are products of culture engaged in a political process. As such, why do we at least partly, and in some cases completely, avoid evolution? What ends does the process of muting evolution serve? The depths and profile of this avoidance are the objects of this book. I will probe the extent and nature of this thing—this *negative dialectical space*—where that which is paradigmatically profound within the interpretive repertoire of the biological sciences is shunned by a slight majority of Americans. A robust rationale for this stunning state of affairs is what we are after.

Anthropology then is the correct lens by which to apprehend this avoidance. What counts regarding one's receptivity toward evolution comes from a holism of life experience and *being-in-the-world*. As a function of culture, we have all been variably and differently socialized to see the ends of *Truth, truths, or no Truth* by epistemologies specific to our local circumstances. Our Western conception of *Truth* is strongly shaped by religion and science in culture. In this way, education then is a process that incorporates narratives from the home, schools, and houses of worship. As an incorporation of experience housed in a body with ideals projected out into the future, we each have ends by which we work. We are, simply, in the world toward ends.

As an anthropology of science education, my ethnographic description of our cultural avoidance of evolution sits nested within a larger meta-discussion which I begin here and rejoin in Chap. 8. Between these bookends is a story of science and religion at play in the field of education. I introduce ideas here which, after the presentation of my ethnographic story, will animate future directions and point toward next steps. Attentive to aesthetics, I ask you to keep always in mind my multiple usages of *evolution*, *end*, and *world*. I also ask you to consider the lyrical epigraphs I will leave for you throughout the work. Selected to amplify mythic American narratives, metaphor, and symbolism, these works of traditional culture, and a few from abroad, resonate with themes and experiences of those within. Chap. 8 will enjoin elements of what I am about to discuss with analytical and summative insight from the main ethnographic description.

1.1   The End of a World

Ends are multivalent things. On one hand, ends are the ideals we have or work toward. Some see our existence as a state of looking out into the world and working toward some final description of all that we can possibly apprehend. For a small amount of others, ends are goals we set and work toward because we see them as solutions to problems. There will be endless more, and they will use the tools of the time to try to improve our lot in the world. For a large group, ends are set by a Godhead whose instructions we best follow lest he smite us. Taken together, ends then are the projections of ontology, our most basic way of talking about our being here in the world. The difficulty of speaking about ontology, for those who have not considered it or care not to, is, as philosopher Hubert Dreyfus puts it, like trying to show a fish the water within which it swims (Dreyfus 1991). For what I am about to discuss, we are the fish whose water is our apprehension of reality through the competing lenses of science, religion, and cultural life generally. These views toward ends, and how these views compete in the discourses of life, are at the core of our problem with evolution. For more than a few, evolution is poison in the water.

Ends also have a darker, but more familiar apocalyptic usage that speaks to this act of poisoning. We can all picture the sandwich-boarded crazy man with a bullhorn prophesying about the end of the world. Although Mr. Sandwich Board is an extreme, the core message of this extremity is rooted back to and undergirds the dominant view of Christian religious practice today. This type of end was not always such a dominant narrative in human history, but it has been so for the West since the rise of Christianity. As Thomas Sheehan (1986) has described it, this type of end enters the Judeo-Christian conceptual world after the Babylonian exile. After returning from Babylon, the mythic narrative structure of Zoroastrian dualism had sunk into the narrative structure of the Judean mythos. Time now had a definite end and a future better world was coming. Political turmoil exacerbated this shift in Jewish theology in ways that reshaped views toward God:

> At the beginning of the Maccabean revolt, when Israel's fate seemed to be at its lowest point, pious Jews began to hope not for a new divine intervention *within* history but for a catastrophic *end* to history, when God would stop the trajectory of Israel's decline by destroying this sinful world and creating a new supratemporal realm where the just would find their eternal reward (p. 39).

As Sheehan sees it, this is the birth of eschatology: a radical reconceptualizing of worldview—a "doctrine of the end of time" (p. 39). Popularized by authors writing in this genre, eschatology becomes known as *apocalypse*, and the cultural template is set for what would come to be a dominant narrative structure within Judeo-Christian theology until today. From the Greek *apokálypsis*, this "unveiling" portends of worlds to come. But with such a concept, we also can consider the worlds of the ancient past. As Darwin and Lyell would synthesize for both biology and geology, ecological worlds come and go. Punctuating the fossil record, mass extinctions mark the ends of geologic periods. Cultural periods and their beliefs about God and nature are no less susceptible to such extinctions.

1.2   End(s) of World(s)

A large number of Judeo-Christians worldwide have found ways to integrate evolutionary theory with a non-literalistic understanding of the Bible. In the USA, a considerable minority, and perhaps a majority, has not (Pew 2005). As sociological changes have taken place, the religious traditions represented by Creationists have begun to seem increasingly odd, out of step, or downright crazy to those from the other side of the modernist looking glass. Perhaps prompted by the increasingly metropolitan-centric discourses of modernity, Creationists continue to look to a form of *Truth* which most intellectuals have simply left behind. Intellectual discourse though is one small circle of all discussion in a society, and those that typify an antagonistic relation toward evolution come from many other walks of life.

Historian Ronald Numbers (2006) has followed the rise of the modern Creationist movement. As he argues, Creationism born within the USA may actually be spreading in pockets around the world rather than withering. Just south of Cincinnati, Ohio, in Petersburg, Kentucky, sits the Answers in Genesis Ministry Creation Museum. Led by Ken Ham, modern Creationism's loudest mouthpiece, this ministry evangelizes an apologetics message that sees the earth as about 6,000 years old. Having come from Australia, Ham leads an institution that now has both the means and the social connections to attempt to spread his ministry's message further worldwide than any Creationist before him.

Within the Creation Museum, the general narrative thrust is one of humankind turning from Biblical authority. As Answers sees it, with the tools of the enlightenment we have turned more and more from God. After Descartes conceived *res cogitans*, science increasingly pointed to humankind, rather than God as the measure of all things. Writing shortly after Darwin, Nietzsche (Nietzsche and Kaufmann 1974) warned that due to the success of science, increasingly tortured religious logics (such as that of Answers) would emerge:

> Looking at nature…as if everything were providential, a hint, designed and ordained for the sake of salvation for the soul—that all is over now, that has man's conscience *against* it, that is considered…indecent and dishonest by every more refined conscience … (p. 306).

Nietzsche saw the enlightenment as opening up new possibilities of being rather than the latest mediator of a divine order. This self-overcoming, as Nietzsche often described the human condition, has a quite evolutionary flavor as Richardson (2004) discusses. New ideas that proved better as new problems emerged supplanting the old, having traction for their ability to free the most people all the while beholden to no one great *Truth*—this was Nietzsche's project.

The corollary of this project, and the point of Nietzsche's (1974) aphorism of the Madman, was that "God is dead"(p. 181). Science for Nietzsche, as first a project aligned with the church in its search to ever more accurately describe the wonders of God's creation, was now so successful that it was seemingly undermining the possibility of God's existence. Although having broad impact within the social sciences and humanities, the implication of the message of God's death fell flat for most natural scientists, as Nietzsche described in 1887, and remains mostly true today. Discussing the faith that some still see in science to provide us with access to one final *Truth*:

> It is still a metaphysical faith upon which our faith in science rests—that even we seekers after knowledge today, we godless anti-metaphysicians still take our fire from a flame lit by a faith that is thousands of years old, that Christian faith which was also the faith of Plato, that God is the truth, that truth is divine (p. 283).

Naturally, this message sounds preposterous, shrill, or perhaps heretical to some. The implication that Nietzsche made—that science had no special access to *Truth*—reduced the endeavor of science from grandly divine epistemological champion to one of an also-ran. For Ken Ham and his Creationist brethren, the implications of the apparent transcendental pointlessness of life as disclosed by evolutionary theory simply cannot be tolerated. Natural scientists who fall into a similar sort of fundamentalist trap of scientism begin to sound equally shrill and out of touch as they *really* believe that the natural sciences will somehow articulate a final theorem of reality.

The move I have just made is usually seen by logical positivists as one of obfuscation. To accommodate "other ways of knowing," which for positivists is usually seen as postmodernist code, is seen as bringing religious inspiration into reason. Although I will be discussing religion, I have no religious faith. My interest in discussing this stems from a desire to find novel ways toward getting evolutionary theory taught. One main message from the world I am about to illustrate is not new—that science is a cultural enterprise—but many natural scientists have been a bit slow on the uptake. This focus will demonstrate that the shrill tone currently proffered by public intellectuals such as Richard Dawkins not only does more harm than good, but is also indicative of his own limited intellectual repertoire toward dealing with the importance of culture. During this project, Ken Ham and Richard Dawkins will then serve as symbolic placeholders—a shorthand for those in society who have similar ideologies. We will also examine those more "agreeable" folks that lay between these caricatures, but as this is so close to home for so many, this will be a more subtle Durkheimian exercise. Ecumenical religious practice in the USA is a normative hegemon—one that resists being analyzed, but from which I will not refrain.

As a product of American culture, shaped by both the trajectory of science and religion, Creationism faces a crisis of sorts. The success of the evolutionary narrative for them represents the end of their world. That evolution represents a fact of coherence matters little to the committed Creationist. At its grandest, the factical "truth" of evolution is the most infuriating to Creationists, as all it offers is a story of never-ending change. How one begins to see possibility in evolution rather than apocalypse is perhaps the description we need.

1.3   Why Did Darwin Hesitate?

Darwin paused. As is well known from Darwin's personal notebooks, the transmutation of species by material means had strong implications for the authority of religious interpretation. That humankind be a difference of degree from the beasts rather than that of a differing kind specially chosen by God?—the implication cut into the bone of nineteenth-century Christianity, questioning a fixity of the natural world until then uncontested by revolutionary challenge. Soon after publication of the *Origin*, public reaction was as strong as Darwin's argument. Enshrined in the mythos of science, Darwin's fiercest advocate T.H. Huxley debated Bishop of Oxford Samuel Wilberforce 6 months after the publication of the *Origin* at the Oxford Natural History Museum. This would be the first of what would come to be a series of public forums by which the social implications of evolutionary theory would be contested. Moved across the Atlantic and 65 years later, the American Scopes Trial of 1925 in Dayton, Tennessee, again crystallized public sentiment toward the notion of evolution. Case by case, public receptivity toward evolution found its way into the spot light of the American legal system during the twentieth century. In 2005, attention focused on the small rural school district of Dover, Pennsylvania, not far from my childhood home. Evolution continues to challenge many to this day. Why?

Like the impact of Jesus Christ reconfiguring a view of the Jewish world, Darwin would have similar impact on Christendom. Like Christ's apostles, and the various ecumenical councils that have articulated what counts as Christian religious *Truth*, evolutionary theory continues to be reshaped by the practices of scientists. As one may choose to see it, we are in an interstitial age where the power of the scientific narrative increasingly becomes omnipresent; it is unclear what Judeo-Christian religious discourse will survive Darwin's apocalypse.

1.4 Why Do We Continue to Hesitate?

In the summer of 2009, as part of the 150th anniversary celebrations marking the publication of the *Origin of Species*, I took part in a conference discussing 150 years of "religious responses to Darwinism" at Oxford University. Amidst the presentations, Ronald Numbers, a celebrated historian of Creationism, stood out. He particularly stood out to me in this instance not for his usual exemplary scholarship, but for a claim he made which underscored a void in the American intellectual understanding of antievolution sentiment. As part of a discussion fearing that the rhetoric of the "new Atheists" movement, that of Richard Dawkins, Daniel Dennett, Christopher Hitchens, and Sam Harris, would somehow prompt US courts to associate science with Atheism, Numbers warned, referring to Dawkins, "He's dangerous…If the US courts listen to this—if they associate evolution with atheism, you won't be able to teach evolution." What immediately struck me as odd was the impression I got that Numbers thought that the teaching of evolution in the USA was business as usual. As you will shortly see, it clearly is not.

Chapter 2
Evolution Education: A Lay of the Land

2.1 Debate!

I first stepped onto the campus of Mason-Dixon University during the fall semester of 2008. I was there to study the ways that students interact with the concept of evolution as they encounter it in university science education. As an anthropologist of science education, I am interested in the ways culture—specifically religious fundamentalist strains within American culture—mediates teaching and learning in science education. My intention on that first trip was to make contacts and build rapport with gatekeepers within the institutional structures of the university: the Dean of Students, student organizational leadership, biology department chair, science education professors, etc. I would familiarize myself with the campus and secure permission to participate as an observer. Rapport with faculty would be built. A gloss on the material and programmatic culture—data from schedule books, student requirements, courses of study, environmental artifacts such as postings, handbills, etc.—would be collected. I would review the campus and regional newspaper archives for historical controversy regarding evolution. I would get a sense of general student behavior and practices in the places I knew I would observe, and get a sense of other places that might be important. I would do all this with the intent of digging a bit and finding where there might be contention regarding evolution. In my second meeting (with a student affairs officer) on my first day on campus, I was handed this (Fig. 2.1).

Unbeknownst to me, but entirely unsurprising, this campus had both a clear tension among the student body and a pending event which crystallized antievolution sentiment in the USA. Why was this institution, a publicly funded university, giving a forum for "equal time" arguments on their campus?

D.E. Long, *Evolution and Religion in American Education: An Ethnography*,
Cultural Studies of Science Education 4, DOI 10.1007/978-94-007-1808-1_2,
© Springer Science+Business Media B.V. 2011

Fig. 2.1 "Mock Trial" advertisement posted around campus

The debate was scheduled for the following evening. I rearranged my schedule and made sure I would be there. In the meantime, I discovered that the biology department, stung by having been only given cursory input as to the direction and content of the program, placed the following public statement on their department web site:

We, the members of the [Mason-Dixon State University] Biological Sciences Department, would like to provide some clarity on an issue that has recently been the topic of conversation on our campus.

Science is a process that involves the observation of natural phenomena, the development of hypotheses and the testing of predictions from these hypotheses through the systematic collection and analysis of data. Conclusions are drawn only after the rigorous application of this process. While all scientific ideas are open to challenge (indeed, evolutionary biologists continue to test these ideas every day), basic evolutionary principles have survived over 150 years of repeated testing. Evolutionary ideas are thus based on observable data rather than belief, and they form the underlying framework for all of modern biology.

Potential alternative explanations for biological change have a high standard to meet, and must survive repeated objective and replicable testing to be considered valid in the scientific context. Creationism, "creation science" and intelligent design are philosophical viewpoints with predetermined conclusions based on belief. They are also inherently untestable. Such ideas therefore constitute unsubstantiated explanations for the diversity of life on earth and cannot be defended on scientific grounds. Introducing such ideas into the biology classroom would be akin to teaching astrology in an astronomy class or alchemy in the chemistry curriculum. Everyone should be free to believe in whatever credo they choose, and we respect those beliefs, but that does not mean it should be taught as science.

We are troubled by any implication that evolution as an underlying concept in modern biology is in question. It is not. These arguments have been settled in a biological sense for many decades, and in a legal sense more recently (Kitzmiller v. Dover Area School District as it regards the Establishment Clause of the First Amendment). The continuing controversy regarding evolution and creationism here, absent in most developed countries, comes from proponents with a specific ideological agenda. In short, at the core of our concern is anything that creates confusion (or provides misinformation) about what constitutes science and how scientific conclusions are reached.

As three different faculty members would later attest, students from the Mason-Dixon College of Law, along with a campus center for public engagement, had approached the department about their input toward potential debate. After this initial meeting, no further contact was made until the program was set and the advertisements were distributed.

The debate commenced the following evening. An equally mixed crowd of faculty from biology (whom I would later interview at length), students, and community members took seats around me. In what, regardless of your opinion of the topic, ended up a confused affair, neither "side" had a clear notion of the intent of the evening. Adjudicating the scientific validity of the points?…the legal status of equal time legislation? Folks with whom I spoke uniformly seemed puzzled. In my own assessment, there was much potential dialogue lost between the competing vocabularies of law and science, of which few people present had a reasonable handle on both. In the student newspaper the following day, the op-editorials voiced much the same.

I was struck by how clearly the statement by the biology department and the fumbled intent of the debate were indicative of the American cultural relationship toward evolution. As the department statement gets right, attempts to overturn evolutionary theory have a very high bar to clear, toward which Creationism and Intelligent Design proponents have yet to make the slightest increment of visible progress. But at the same time, the notion that Creationists are somehow driven by ideological ends which work-a-day scientists stand outside is laughable and at odds with what is at stake in taking evolution education seriously. The department statement starts off by deflating science as a process. A process—something that to me sounds more akin to making hot dogs or widgets—certainly does not capture the nature and status of capital-S-Science in American society. I do not fault many practicing scientists for being shortsighted on this one. Many scientists I have encountered demonstrate disdain when considering why some students simply shrug or show disinterest at all or part of the knowledge the sciences produce. This disdain manifests itself with particular focus when directed toward religious fundamentalists. But more widely and successfully presenting evolution education, respective of the issues that religious fundamentalists bring to the civic discourse, will require tools and concepts outside the usual purview of most scientists and science educators.

2.2   My Perspective

Why do I care about this, and what prepares me to address evolution education ethnographically? This story started very early. My mother has repeatedly, during critical moments in conversations, warned me that I am going to hell. Asking provocative questions—like the basis of our family's religious faith—violated a quartered off area of thought that she had not apparently ever considered exploring. For her, my intellectual pathway was tantamount to taunting God. You see, similar to many I know without religious faith, I came to this position through my coming to understand biological evolution. This was not for my parents' lack of trying to socialize me toward friendlier and cuddlier positions—I made matchstick crosses and God's eyes with the best of them at summer Bible School. One of the proudest narratives my mother tells involves a church elder commenting that my earnestness and precocious manner might see me take up the mantle of pastor one day.

Although mostly confusing to them, my theological journey really is not that profound. Like many Americans, our family's theological bar was both set pretty low and was rather flimsily constructed. When the questions I brought toward religion were met with what at the time seemed like secretive dodges, I now understand that the adults of my life did not know what to do nor how to answer. Readers and contemplators these people were mostly not. Working with only the slightest theological knowledge, but deeply called to do the appropriate moral duty of raising children *as one does*, off to church we went. Having "arrived" as middle class in rural Pennsylvania through generational investment of working-class migrant stock, there was little enquiry regarding other ways of life. Life was good, and God had blessed us.

In fourth grade, I discovered through a library book, that humans had apparently evolved along with all other animals on earth. At the same time, my closest childhood friend—who happened to also be a religious fundamentalist—would argue like cats and dogs about science, faith, and what counted as real. Think fourth-grade metaphysics. In lieu of any intervening theological influence (my parents had a very small social network), I was left to adjudicate science and faith on my own. Tangible faith practices around me were almost entirely Protestant Christian, accentuated by the ascetic influence of the Amish, Mennonite, and Brethren religious traditions—novel elsewhere but distinctive in my childhood environs. In the end, faith did not put up much of a fight. Nor were science and religion ever presented as potential dance partners. Offering dusty tales of Levantine political struggles, religion faded while the shinier rocket-fueled promise of science beckoned and won. Science could speak to me in clear terms through the pages of books right there on the library shelf. Faith seemed unnecessarily opaque and was translated through fumbling Sunday school education. The contradictions, the glaring problem of other faiths, the "problem of evil," and the moral absolutes—it all seemed to confusing to me as a child.

Although the remainder of my schooling was marked by my general contempt for most everyone I encountered, there were moments of respite. A handful of teachers and peers were genuinely interested in what I have now come to understand as the life of the mind. This made those years tolerable, and unlike those who find the romanticism of schooling to be the "good days," the prospect of going back nauseates me. At the same time, I escaped into the world of teen after-school and weekend work, cooking at a local nursing home. Inside I honed some of my more contemptuous thoughts regarding some of our social values as I watched the residents go about the small routine circles of a slow, quiet death—in practice mostly abandoned by their families. Confronted what I saw to be a contrived social existence disconnected from the richness that constituted one's prior community content, I saw residents' spirits fade into shells of what once must have been. Death hung with me like an unshakable atmosphere.

I eventually left home as a first-generation university student. College opened up a much broader acceptability for my dispositions. My focus on effective science education came from a grounding in multiple areas of study; anthropology, geology, philosophy, science education, and religion all led me toward eventually undertaking the research for this book. Like any good liberal arts education, this repertoire gave me a more mature understanding of science, faith, and other people in the world. After some years doing archeological fieldwork, I entered the science classroom with the intent to improve the quality of evolution education.

The experience of craft knowledge, I believe crucial to anyone attempting to do educational research with impact, reintroduced me to the nature of the education business. The bureaucracy was not surprising. The depth and extent of the anti-intellectualism was. When I aired my research interests with my colleagues, their reactions might have suggested that I had just violated a social taboo. In my working-professional master's degree program at the time, the science educator recoiled in a warning to me about pushing the evolution issue. In my own classroom, at the very slight mention of human evolution in a side discussion one day, a sixth-grade student, as if struck through by a higher power, jumped to his feet and commanded "I didn't come from no stinkin' monkey!" It quickly became clear to me that the actual practices regarding evolution in the classroom were not quite as "on the ball" as I might have hoped. With this in mind, I knew like many that something was not quite connecting between our curricular ideals and actual practices in science education.

2.3 Evolution Education in the USA

> My husband and I think that aliens may have been involved in evolution ... I mean, all that stuff you see on the History Channel about the Pyramids and Stonehenge and how they couldn't have built those things by themselves ... how do they know?
>
> A College of Education administrator's
> response to the gist of my work.

2.3.1 Rethinking Our Approach

To understand why an ethnography of evolution education is necessary depends in large part on what you see counting in educational dialogue about how one learns.

If science knowledge and learning were a matter of being shown discrete points of *Truth* and adding these bits to our respective *Truth*-piles, there simply would be no societal issue over evolution. Rather than one having misconceptions regarding this or that contentious bit, it is more honest to acknowledge that knowledge is political—often mundane—but nonetheless political. Those who reject evolutionary theory do so not out of malicious ignorance, but as I will show, out of having no tractable use for it in how they organize the guiding framework of their lives.

Most education research regarding the American relationship toward evolution has yet to take the foundational framework of being (student ontology), and what it prescribes for some learning seriously. For Creationists committed to a literal reading of the Bible and what that dictates about knowledge, some scientific knowledge simply will not work within this framework. To only argue that such people have misconceptions (although *very* tempting), is to miss the central organizing principle that animates some lives—the role of religion in shaping reality.

All research variably engages with explicit and implicit connections to theoretical perspectives. Some, like social movements or tastes, dominate the ways we think about things, and some push the envelope. Some researchers, passive in their deference to perceived giants of their fields, do not seriously interrogate the epistemological legacy and impact of knowledge systems within which they have been socialized. As Kincheloe and Tobin (2009) have pointed out, even though there has been a general move toward more heterogeneous theoretical approaches in science education, there still remains a great zombie-like albatross of logical positivism ever-circling around and in some cases roosting within the field (even though pure logical positivism has, for the most part, long ago died in philosophical circles). As an anthropologist whose gaze has centered on science education, I am naturally interested in and strive to be aware of the organizing principles (so much as they exist in apprehensible form) that various actors in social fields bring to the various nodes of social discourse—in this case evolution education in and out of the classroom.

Educational anthropologist Hervé Varenne (2007) has asked simply "where do people learn about Creationism and Intelligent Design?" (p. 1562), within a piece that discusses what he describes as the difficult collective deliberations which take place regarding school. Often, these deliberations consider the content and politics of schooling, but due to the tensions that such discussions involve, take place outside of school. Evolution education takes place under such a social arrangement, as demonstrated throughout this book. Adding perspective to this, Fensham (2009) has noted that science educators often greatly over-project the importance and impact of their work, as if once a research claim is made, it represents some form of immutable *Truth* which must be paid tribute. There are many corners of *Truth* that most will never encounter nor will ever garner interest.

This work broadens the discourse of evolution education research to a more robust consideration of the socialization involved in schooling, and whose influence counts regarding evolution. Such an analytical move takes us outside of the influence of the classroom by considering the relative power of all discourse in a student's life. By politicizing the actions and discourses of teachers, students, administrators, and parents, we then begin questioning the extent to which the normative practices of those walking into and working in the classroom are in fact "the problem."

To continue to scratch our heads at why many Americans have little interest in science and a large minority reject scientific orthodoxy is to simply ignore all the ways and places one learns, or more importantly finds significance in their lives. To make this mistake is to perceive science education in total as an act of transmission—timeless objective *Truth* copied and moved from point A to B within the four walls of classrooms and laboratories. Perhaps as Apple (1993) put it best, "it is naïve to think of the curriculum as neutral knowledge. Rather, what counts as legitimate knowledge is the result of complex power relations and struggles among identifiable class, race, gender, and religious groups" (p. 46). So how can one discuss evolution education without discussing the effects of religious education, social class, or the contingency of local circumstances?

I will begin by identifying the groups for whom evolution becomes a "problem," and how these people interact within schooling to culturally produce antievolutionary attitudes. Along with these folks, one can also see those for whom evolution cues ambivalence, and those who act with as much zeal for evolution as that of Creationists who act against it. How these groups interact in the practices of schooling then becomes a dialectical performance which I will illustrate. This illustration then will provide a clearer picture of the phenomenon of evolution education. As I will show, resistance to evolution is resistance to a symbolic construct—one of a social world of human practices (people doing science) which most students either could largely care less about or specifically reject parts of. Try as we might, as some students are brought into an understanding of evolution, the social structures that produce Creationist dispositions continue to churn out more of them.

Rather than treating those who object to evolution as static objects, what rationale based in the contingencies of cultural logics borne of distinctive social worlds can we explore? Rather than assume a Creationist has "misconceptions" or concepts which need changes—both models based in a deficiency view of education rejected by many in the field—we will suspend judgment for now to more deeply understand the phenomenon of rejecting evolution. The deficiency model of education essentially amounts to blame the victim approach to pedagogy. Bernstein (1971) and Valencia (2010) are key works critiquing this view. More humanely, what in fact do Creationists *have* that makes the study of evolution troublesome or even dangerous? Extending through the social field of all of education, how and why is it that so many American teachers avoid teaching evolution? Where did they "learn" to do this? What purpose does this avoidance serve? How extensive is this leveling down of possible educative dialogue regarding evolution? Although I will not attempt to address all these questions, this is the general lens through which I will illustrate. To get a handle on what is at stake, let us think about a quick concrete example. Consider a student who spends 3 hours a week for a semester being talked at by a biology professor who only occasionally mentions evolution. At the same time, this same student has spent 20 years of their life building lifelong familial and community bonds with their country pastor who knows and professes that evolution is "the work of the Devil." There is certainly going to be some conflict. As I will show, for the student this conflict manifests as a strong but internal dialogue—difficult to measure in the positivist sense, but rather easily illustrated once you become an instrument and build rapport with people.

What exactly then is going on when someone says they do not "get" evolution? When they assert that "there's not enough evidence," or that "some other" science exists that critically undermines evolutionary theory? As I stated at the onset of this book and reiterate here—the phenomenon of resisting evolution is not a "scientific" one. That is, the nature of this resistance is not a discrete object of scrutiny by which the tools of the natural sciences will be ever able fully, finally, measure or control on their own.

As I am interested in how people construct their understanding of evolution and how ideology mediates perception, I proceed in this book informed heavily but not exclusively by the existential/phenomenological theoretical tradition. As a bricoleur, I also will pull complimentary perspectives from various threads of sociology—cultural, religious, and political economy. For my central purpose, the methodology of phenomenology orients an investigator toward carefully listening to and attempting to represent the constructed world and cultural commitments of those with whom you investigate (i.e., I will take Creationist commitments and understandings of the world at face value). For the purposes of researching breakdowns (the where, how, and why we fail) in evolution education, this proves especially salient. In this spirit, the following list of common evaluative judgments toward Creationists are the types of things which I will be setting aside:

1. Creationists are stupid.
2. Creationists have misconceptions about science.
3. Creationists are antiscience.

Why is it important to set these aside? Am I being "soft on Creationism" as Toumey (2004) encountered in the public critique of his ethnographic work with Creationists Engineers in North Carolina? No. Toumey's interpretivist writing both was fairer to the lives of Creationists, while refraining from pandering to the polemicists in the sciences; all the while agreeing with orthodox science's conclusions regarding evolution. I continue in Toumey's tradition, but expand the interpretive mission to not only make Creationism more familiar (given its impact within education), and make normative religious practice strange (given the paucity of theological literacy I will present). As I will make abundantly clear, the nature of what is at stake for Creationists as they dismiss evolutionary theory has not yet been clearly articulated by the scientific or science education community.

Additionally, past the *seeming* irrationality of Creationists, the broad middle of American civic discourse has a surprisingly high tolerance for any reasonable position that comes to the table of dialogue. As I will show, the "teach all the theories" strategy of tweaking the science curriculum currently plied by Creationists has wide but nonetheless superficial appeal to most Americans. This superficiality is not something easily overcome. As I will illustrate, the tension for some between science and religion in the USA is built on quite shallow understandings of both.

Politically, the fact that such Creationist strategies are simply indicators of a broader agenda of advancing fundamentalist theocracy within the American public school system is not recognized by any appreciably large group.

Within the ethnographic data that make up this book, for judgments one and two above, there are simply no clear grounds on which Creationist students stand out from other students that warrant such labels or assessments. Anyone can be made to appear stupid when measured against an area of knowledge for which they hold no vested interest nor expertise. The fact that an area of knowledge describes something "real" says nothing about public receptivity toward it. On these grounds, the logical positivism that dominates much of science education pedagogy becomes a form of cultural naiveté. As I will discuss regarding judgment three, Creationists are often just as likely to be science majors as not, and often are quite competent at passing classes.

2.3.2 Just How Bad Is the Status of Evolution Education?

Historically, evolution has long been contentious in the American educational enterprise. Although perhaps counterintuitive, the publication of *The Origin of Species* did not immediately cause uproar in the American educational system. Numbers (1998) describes the reception to evolution within the scientific community as largely settled, barring the odd holdout, by the end of the nineteenth century. The public's contention with evolution coalesces and then comes to the foreground in the years just leading up to the penultimate stage for evolution—the Scopes Trial of 1925.

At the time, and in the next few decades, resistance to evolution grew along with participation rates of public schooling. Until American schooling becomes compulsory during the twentieth century, large numbers of people remained out of contact with more "progressive" forms of education, and the content of its Darwinian-influenced scientific pedagogy. Moreover, institutions of higher education, more so than today, served a more homogenous population that tended to sample from the economic elite. Whereas higher education was once preparation for the manners and practices of genteel status, this has slowly given way to the K-16 system of increasingly compulsory education through university, which is emerging today.

The Cold War fueled space race of the 1950s and 1960s generated a national movement toward a more rigorous scientific curriculum, after the launch of the Soviet Sputnik satellite. In a first broad move toward a national inclusion of evolution in public school curricula, the Biological Sciences Curriculum Study (Hassard 2005) and its prominent inclusion of evolution in its textbooks prodded populist sentiment against evolution once more. Although the content of evolution was now more prevalent than ever, school boards needed to adopt these textbooks, and teachers then needed to teach the concept—both actions that were smattered at best in their national practice. Skoog (1979, 1984, 2005) has traced the poor presence of evolution in general and human evolution particularly in textbooks throughout the twentieth century. Since the 1950s, development of the Biological Sciences Curriculum Study, evolution has been pushed further into the foreground of school curricular discourse, and local communities and organizations have sprung up in resistance to it.

As Numbers (2006) has most clearly shown, the rise of the contemporary Creationist movement in the USA really gets widespread traction in affecting school curricular policy with the publication of Morris's (1974) *Scientific Creationism*, by the legitimacy a Creationist "textbook" offered the movement. Part of a long (and very rich) history of American evangelical Christian theology, from the Great Awakenings through the populism of William Jennings Bryan and then Billy Graham up until the current iteration—the Mega-church—evangelical Christianity (barring a tiny minority of more theologically liberal evangelicals) has rejected evolution. After *Scientific Creationism*, as Larson (1985) showed, a long slog of court battles has ensued first legalizing the teaching of evolution, then outlawing "equal time" for Creationism, and slowly mutating into the more metaphysically oriented "Intelligent Design" debates currently active today. Intelligent Design theory, in both legal cases and philosophical intent, as evident in the *Kitzmiller v. Dover* "Intelligent Design" case (United States 2005) aims to draw on experience and *Truth* outside the traditional parameters of liberal academic practice. By doing so, the aims of this "repackaged" Creationism (Forrest and Gross 2004) seek an ideological shift in the liberal consensus toward evolution—one, which if enacted, is committed to the removal of evolution in the content of curriculum.

In *Kitzmiller v. Dover*, the National Center for Science Education's research team uncovered seemingly damning evidence that the motivating force behind Creationism's most recent face, "Intelligent Design Theory," was simply a political tactic toward an evangelical Christian takeover of science (Forrest and Gross 2004). As the testimony in *Kitzmiller* showed, the Discovery Institute—a leading Intelligent Design strategizing organization—formulated a Christian restoration movement strategy titled the "Wedge." In light of the tactical language of the "Wedge Document," there is no reason to assume more ecumenical motives behind Creationists or their Intelligent Design allies. As Philip Johnson, a noted Darwin critic and Intelligent Design activist, has framed it "our strategy has been to change the subject a bit so that we can get the issue of intelligent design, which really means the reality of God, before the academic world and into the schools" (Discovery Institute 1999). More to the point, Johnson cuts to the issue: "this isn't really, and never has been a debate about science. It's about religion and philosophy" (Forrest 2007).

More extreme versions of Creationism, such as that proffered by the Answers in Genesis Ministries through the public education and outreach strategy of their Creation Museum, attempt to sway public opinion by presenting spurious "Creation Science." This is not to be unfair toward Creationism. A tour through the Creation Museum is fairly straightforward in its social agenda—the restoration of a mythical pan-Christian America where the ills of society, stemming from evolution, are enumerated and attributed to a materialist worldview. Creationist tropes (flood geology, lacking transitional forms) are portrayed for a self-selected audience that largely buys the presentation.

For Creationists, evolution then is simply a straw man or placeholder for a larger and much deeper conflict of ideologies or worldviews. The philosophical/procedural starting point for science today—methodological naturalism—is off-limits as a conceptually agreeable framework for many Americans. It is not simply a matter of ignorance or lack of exposure—rather, some people simply refuse to "go there." A natural next question, given the ideological intent of Creationists, is the degree to which those who teach within the American system of education are so aligned.

As a read of teacher survey data shows, and I will contextualize through people's lives in this project, few teachers take a strong stand for evolution and most simply gloss over it. An exemplary few teach it well. Of the measures of receptivity toward evolution education that have been conducted with students and teachers, the picture is bleak for those who presume that knowledge represents apolitical *Truth*. On these grounds, the policy recommendations such as Project 2061 of the American Association for the Advancement for the Advancement of Science are far, far from even coming close to being achieved.

Secondary students unsurprisingly appear to have conceptual difficulty in avoiding teleological ideas of the origins of acquired characteristics, paralleling a difficulty in their use of genetic knowledge more generally (Clough and Wood-Robinson 1985a, b). While taking a formal assessment of biological knowledge, students misinterpret the meaning of adaptation, often giving adaptation a clear link to a metaphysical purpose (Lucas 1971). As Hallden (1988) describes it, students often respond from a subconscious meta-cognitive standpoint, bringing in justification and points of reason from far outside the immediate hermeneutic circle designed for the class activity. Most pointedly Ervin (2003) considers the impact of high school students' ontological positions when encountering human origins in the biology curriculum. As he sees it, students who are not open to a conceptual change model of knowledge regarding human origins may successfully use or manipulate concepts using what he describes as "plagiaristic" knowledge, but seemingly have not "learned" as Ervin sees the process. For Ervin's cognitivist approach, to have learned, one must have moved a concept from apprehension to comprehension. Translating this point to an ethnographic approach, I will demonstrate the culturally situated rationales by which this happens.

Within university settings, additional nuance emerges that illustrates the pervasiveness of both Creationist views and tolerance of them by students. Of the large number of such surveys that have been conducted, the following are indicative. At Bowling Green State University, 442 students, most undergraduates in the last year of a teacher training program responded as follows in surveys (Bergman 1979):

- 91% responded that both evolution and creationism should be taught in schools.
- 72% of the graduate students also felt that creationism should be taught.
- 6% of the students thought evolution alone should be taught.

Fuerst (1984) conducted a survey of 2,387 biology, anthropology, linguistics, and genetics students at Ohio State:

- 80% thought that "other views" should be taught when evolution is taught.
- 25% reported that they thought scientists feel that evolution is invalid.
- 22% responded that teaching evolution led to decay in society.

Zimmerman (1987) replicates Fuerst's study with Oberlin College students:

- 89% responded as believing in evolution.
- 56% responded that creationism should be taught in public schools.
- 89% responded that evolution was taught in their high schools.
- 6% responded that teaching evolution led to a decay in society.

For Hahn (2005), students that self-identified as Creationist strongly identified evolution as part of a system of moral decay, where results with more secularly minded made no such associations regarding evolution and morality. Dickerson (2003) came to a somewhat eye-opening conclusion in a survey of United Methodist ministers, preservice science teachers, and preservice language arts teachers. The ministers reported the greatest depth of knowledge toward the concepts of natural selection and evolution and the strongest need for more science literacy in the American public, followed by the language arts/social studies teacher education students, and lastly by those who might soon be science teachers.

Science teachers similarly reported variable commitment toward teaching evolution. In Texas, Bilica (2001) finds low-level emphasis in teaching evolution generally, but certain topics such as human evolution were almost completely ignored. She concludes that the presence of antievolution forces in communities still have great effect, as have been found historically throughout a good deal of the USA. Shankar and Skoog (1993) report the majority do teach evolution, but at a very low level of concentration. The most common correlative for not teaching evolution is a strong personal commitment to "religious conservatism." They found almost 30% of teachers in Texas included Creationism in the curriculum. Not to think that this is a purely Southern US phenomenon, Moore (2007) found in Minnesota that one in five science teachers report voluntarily teaching Creationism on an equal footing.

Jorstad (2002) finds relationships similar to the previous studies, but adds that somewhere between 10% and 30% of teachers in Arizona are skeptical of evolution's scientific validity, while 14% overtly emphasize religious explanations for the descent and diversification of life in class. Pushing a bit further into a context-rich discussion, Kyzer (2004) found that of three teachers in Alabama, the depth of embeddedness in their communities was an important variable for their willingness to teach evolution. Eglin's (1984) conclusions are humbling regarding evolution education:

- 16% of teachers felt that Creation science is "just as scientific" as evolution.
- 27% of teachers taught both evolution and creationism.
- 20% responded that local districts provided Creationist textbooks for class instruction (although not required by law).

- Although most teachers who taught Creationism did so by personal conviction, teachers "on the fence" were "reluctant to refuse to teach Creationism if required by the school district, even if they personally disapproved of it."

Additionally, Eglin finds that many teachers saw no problem in simply omitting evolution from the curriculum so as to avoid controversy. Eglin concludes that "if there is to be a concerted and sustained opposition to the inclusion of creation science in the public school curriculum, it will not…come from high school science teachers" (p. 117).

In an Ohio survey of over 1,000 teachers, Zimmerman (1987) finds:

- 18% of public school teachers reported teaching creationism.
- 34% responded that creationism should be taught in public schools.
- 15% reported teaching creationism "in a favorable light."

More broadly across the USA, Affanato's (1986) polling data of 466 biology teachers nationally finds that 5% thought evolution ought to be excluded from the curriculum, 45% thought Creationism ought also be taught. Additionally, almost 20% of these respondents indicated that a high school biology class could be effectively taught without mentioning evolution, and 15% reported evolution and Creationism as having equal scientific validity.

Naturally, teacher attitudes might then translate into variable commitments regarding evolution's presence in state curriculum policy. Lerner (2000) and, more recently, Gross et al. (2005) have examined the state of curricular standards regarding evolution across the USA. Nationally, Lerner found standards regarding evolution to be poor generally, with a few states standing out. Regardless of the general language toward evolution, human evolution remains a more avoided topic nationally. Outside the frame of the psychological/behaviorist models dominant within science education research, this low level of commitment in curricular policy environments indicates that effective evolution education depends as much, or more, on a political milieu willing to construct and enact an evolution-rich biology curriculum. Ingersoll (2006) further complicates this problem by pointing out the low retention rates of "highly qualified" teachers within the field once placed. The majority of qualified science teachers often stay 3 years or less before leaving the field of teaching voluntarily. Often, these leavers are the most qualified and highest performers in their content-area understanding and science-related standardized test criteria.

2.3.3   A Better Representation of What Is Going On

Within science education research, there are comparatively few perspectives regarding how student values and worldview—a product of their ontology—mediate their receptivity toward evolution. Lederman's (2007) open-ended remarks, concluding his review of research on student relationships toward nature of science pedagogies, ask "what is the influence of one's worldview on conceptions of nature

Fig. 2.2 Maybe not the best
sell of the scientific life

of science?" To step fully into such lines of thinking opens a Pandora's box of epistemological problems for those who remain naïvely committed to a monosemic view of ways we can apprehend reality. For most anthropologists though, the object of our inquiry usually prohibits strong adherence to such commitments. I direct this specifically at individuals for whom a rejection of scientific knowledge by students represents an essential stupidity. The simple reality concerning our social existence is that people value many things in life: sports, religion, sex, visual arts, shopping at the mall, music—things given their meaning and significance within the culture they reside.

The importance of this is clearer when we consider that the practice of science itself is a culture. Scientists and science educators are socialized into a world with norms, values, and taboos. For some students, what is at the root of their sometimes nutty views about the corpus of science is a detachment from and general disdain for what they have come to hold as the lifestyle and cultural position of scientific practice. Dinosaurs they like, the social identity of the scientists they see for whom this advertisement was geared they do not (Fig. 2.2).

For tackling antievolution attitudes within science education, there are approaches that are clearly off the mark. "Misconceptions research" and its less-offensive kin, "conceptual change theory," are both based on a fundamental assumption of a computer-model theory of mind. Although this is not a psychological discussion, I do agree with both Dreyfus (1972) and Lakoff and Johnson (1980), that in the end, whatever neurological model gets us closer toward how the brain works and what that means for a philosophy of mind, it must account for the contextual nuance and metaphorical inclinations of human *being-in-the-world*. With evermore detail, natural science can show the interrelatedness and cause–effect relationships between the materials of the universe. Short of having an understanding of the total neurological operations within the brain and how this could ever translate into decoding symbolic meaning, the understanding of scientific concepts remains an issue of education in culture. Whether one comes to learn scientific concepts on their own or in social settings like home or school, there is an essential social dimension to learning for which natural science is only part of our understanding.

An absolutely crucial issue to which we must attend, if there is to be any substantive improvement in evolution education, is to better comprehend the basis on which people detach from evolution and what conditions move them toward it. To contend with this, we need ask ourselves whether the current dominant research paradigm adequately gets at the issue. Conceptual change theory gets it partly right to the extent that students are inclined to go in a scientific direction (Cobern 1996), but this research agenda is still entrapped in a cognitivist framework working toward universalistic epistemology of Platonic ideals. The best examples of the limits of this thinking are demonstrated by Dreyfus (1972) in his critiques of artificial intelligence. The computer model of mind fails when it has to account for contextual differences. Deathly important, what conceptual change theory misses is the role of the "errant" knowledge in the conceptual ecology of a person's experience, which it hopes to replace. What we need to attend to is the importance of how evolution for some appears to the contextual whole—the gestalt of their experience—and how new information is or is not accommodated within this. With knee-jerk ferocity, traditionally dominant voices in the field of science education have struggled to account for the values in the holism of an individual's contextualized life experience when they defy policy intent and turn away from the vocabulary of science.

2.4 The World of Science and the Worlds of People

As this book will detail, understanding rejection of evolution requires us to attend to the variable epistemological commitments that originate from ontology—which are imbued through culture. As shorthand, worldview is a way to do this. The concept of world, and its embodied form worldview, has a long history in the work of philosophers and social scientists. Critical to the work of German philosophy and its influence on social theory, this *Weltanschauung* describes the ideology and perspective of a person in place, connected to and shaped by the sociocultural practices endemic to that place. A Platonic, universal *Truth* or essence it is not. Naugle (2002) has written extensively on the history and development of the concept. As Dreyfus (1991) sees it, Wittgenstein saw worldview as an ontological problem of describing the background practices of a person's life, which he thought could not be worked out, but which Heidegger's (1962 [1927]) *Being and Time* does. More recently, Bourdieu (1977), greatly influenced by Heidegger, uses the concept of *habitus* to describe the relative interests of the embodied person structured by contingencies of the local social world. The view from this *habitus* might effectively be called worldview. Additionally, Habermas (1989) worked in the Husserlian tradition to stress the importance of the everyday life world, or *lebenswelt* for social analysis and bringing about idealized democratic dialogue. Aerts et al. (1994) further considers the problems of constructing a worldview in light of an increasingly multicultural and interconnected world.

I proceed employing the Heideggerian conception of world that moves us past the Baconian assertion that we project significance onto a universe of meaningless and indifferent objects. As I see it, the existential/phenomenological conceptual toolbox of Heidegger and a handful of others helps us clarify that the significance of our world is actually quite primary, and the scientific gaze tertiary, for most of our everyday being. As Wrathall (2006) frames this:

> In this day and age, no one could really believe that we could discover the nature of physical reality—for example, the properties of electrons and quarks and such—by exploring our everyday understanding of things. But there is a legitimate sense in which we use the word 'world' to name something quite different, something like a particular style of organizing our activities and relations with the things and people around us. The world understood in this way simply doesn't lend itself to be studied using the methodology of the physical sciences...Heidegger thinks that the world, understood in this sense, is a genuine phenomenon in its own right, and can't be reduced to a mere collection of physical objects (pp. 20–21).

Wrathall's description is the type of space within which I frame my narrative about evolution education.

I make this worldly distinction against a two-toned academic backdrop that Snow (1962) diagnosed over a half century ago. Like the Ishihara color blindness test, there are some within the academic community who appear to be *culture blind*. The past two decades in the history of science has seen the flare-up of the "Science Wars," whose primary issues are exemplified in the famous "Sokal hoax." But against the foolishness of ejecting a baby with its bathwater, some problems for science *are* best tackled by dialogues within and about culture, for which resistance to evolution is a case.

Is this focus on worldview *really* necessary? Consider the outcomes when the "pure rationality" of science confronts even the slightest consideration of a cultural construct as important to informing effective science education strategies. In the fall of 2008, champion of evolution Richard Dawkins played along with a reactionary ouster of Michael Reiss, education director for the British Royal Society. In a presentation to the British Association for the Advancement of Science, Reiss suggested that his years of teaching had taught him that for Creationists, little insight from evolutionary theory would change their minds. Thus, he suggested that addressing worldviews, rather than misconceptions, might be an effective teaching tool when teaching evolution—toward which some members of the Royal Society took umbrage. Upon also finding that Reiss is an ordained Anglican minister in addition to being a biologist, society members and Nobel Laureates Richard Roberts, Harry Kroto, and John Sulston each signed a letter to the society president in which they questioned whether Reiss being clergy compromised his reasoning, of which Richard Dawkins sarcastically likened the situation to a Monty Python skit (Ahuja 2009). Although certainly not Reiss's intent to prompt such a response, vocal members of the Royal Society likened this worldview talk as naïve and possibly dangerous. Shortly after the next session of the Society, Reiss resigned. Worldview, for Nobel Laureates, was not going to fly.

2.5   Worldview and the World of Science Education Research

Within the professional literature of science education, worldview has gained greater attention in recent times. In a lead paper of an edited volume that takes up the proposition, Hugh Gauch considers the American Association for the Advancement of Science's (AAAS) "pillars of science" and systematically analyzes arguments as to whether science has a specific worldview (2009). Although Gauch's paper and his interlocutors stay mainly within a positivist framework, other science educationalists consider drawing a wider circle. Cobern's (1996) extensive work on worldview and science education also informs my perspective. Cobern (2000) makes some important distinctions:

> Only an imbecile or an incredibly biased person would reject what are clearly the facts. It frequently happens, however, that students do not accept a concept as the teacher thinks they should, and for this disturbing situation one of two differing conclusions is offered. A teacher will conclude that a student simply does not understand the presentation of the facts, and therefore more explanation and better explication is needed. The problem, in short, is a problem with communication. Science education researchers, sometimes on the other hand, conclude that the problem is rationality. The student does not comprehend, and therefore is not a formal thinker. In both cases the teacher as a figure of authority maintains a posture of empirical chauvinism (pp. 232–233).

Naturally, most attempts at proffering solutions to resistance to evolution have followed along Cobern's two identified tracks—the application of more reasoned argument and the employment of better descriptions, both of which serve the ends of "conceptual change." Joining the science education literature, this book illustrates evolution education in a "world" which one can describe.

As an introductory primer, let us ask ourselves the question—how does "one" learn about evolution? In its essence, let us think about the deconstructed parts of the question. Using Heidegger's (1962 [1927]) concept of *das man* or "the one," how does "one" learn about evolution? This is "the one" of normative practice, what "one" does when wishing to resist the anxiety of stepping out of the flow of everyday, "transparent practices" (Dreyfus 1991). For example, thinking this way, teachers who shy away from teaching evolution, due to the social pressures within the community they teach, seem to have a contextually sensible stance—they have learned to cut the stress of conflict. For reasons that range from fear of alienating student religious belief, to skepticism of scientific evidence, in the normative practice of most education "one" largely ignores or downplays evolution in the American school, lest you bring angry parents, colleagues, administrators, and zealous community members against you. Both Wittgenstein (1953) and Kierkegaard (1962) described this as the process of "leveling," the social equi valent of a kind of lowest common denominator toward socially risky or salient behavior. The most important part of what I sense to be the stumbling block concerning evolution education lies within a rich phenomenon associated with this distinction.

The next question, given this banal teaching of evolution is: what does "one" learn about evolution? The answer to this question then is a description of the output of the practices that teachers actually engage in, often doing evolution short shrift in the classroom. But what we are after here is a gauge of this—the nature and extent of this phenomenon, and the contexts in which this happens. The problem for our positivists is "learning" to do this is often nonverbal. This is entirely missed by conceptual change theorists, who chauvinistically presume everyone *does* best practices.

Within this world, evolution only becomes knowledge when used in "transparent practices" (Dreyfus 1991) in a worldview congruent with the constitutive terms for which evolution works as a concept, that is, an ancient earth, the possibility of abiogenesis, and the biological evolution of humankind. For example, this person will have long since stopped thinking at all about the contentious nature of evolutionary theory— it has become natural to their *being-in-the-world*. Naysayers might feel I am exaggerating the issue. I am not and if you think so, you have not looked. Caught in an individualist rationality, they will report an excellent example of evolution education from a "good" school, but sadly this is far from the norm. In fact, the culturally imbued significance of how one signals a "good" school is part of the object of my analysis.

2.6 Studying the Total Phenomenon of Evolution Education

Quite simply, this project begins at the end of evolution education. As an ethnographic study, I followed a group of college students as they encountered evolutionary theory in their introductory biology classes. For many of the students whose voices I will relay to you, this was for them the last time that evolution will likely be discussed in a formal educational setting. Naturally, biology majors will continue on in this process, but for most, the content of science education will begin to fade. As a matter of practicality, given the contentious nature of evolution in American life, starting at the end allowed me an easy way to track back into a student's life, and examine the varying success of science educations ideals as they actually play out.

My interest centered on public university science classrooms, as much better suited than most K-12 settings, to offer phenomenological insight into the practices that typify the American relationship toward evolution. In the public university setting I could expect, more so than in K-12 classrooms, that evolution will in fact be taught. Moving forward with this worldview framework, let us now finish the job and fully embed science as a cultural enterprise, made up by people with perspectives at best respective of a naïve realism, at worst susceptible to ideologies as strident as Creationism.

2.6.1 Starting at College and Going Back Through Educational Lives

There is a saccharine metaphor of educational possibility throughout the well-known Public Broadcasting Service higher education film *Declining by Degrees*. Against alleged trends of grade inflation, overreliance on the marketing of athletics programs to bolster financial support, and a student body engaged in a detached détente from faculty, college presidents, policy makers, chancellors, faculty, and writers line up in the film, almost misty-eyed, extolling the highest virtue of college. Somehow, during the process of higher education, "something magical happens". This may have been the case for a few (perhaps those who became academics themselves?). But hyperbole aside, across all the forms of higher education, from community and technical college to elite campuses, often little *magic* happens. A lot of credentialing happens, the "exchange value of education" as Labaree (1997) puts it best.

Contrasting this, complicating factors abound for students—financial difficulty accompanies increasing costs and the competing expectations of different spheres of socialization. In this, each social "world" has its own vocabularies and outlooks which compete for time and your allegiance, and may steer one's discourse toward silence. Although it is oddly jarring to think this way—what does your fraternity or sorority have to say about evolution?...how about your basketball team and their fans?...what about the Campus Crusade for Christ?...the College Republicans? One might have naïvely expected there to be little or no resistance to evolution, this being a university after all. But each competing field of student interest has their own vocabulary, expectations, and rights and rituals—most of which have little to do with science, some of which works directly against it. Students drop out. They cannot afford the time and money. They lose hope in formal education as a redemptive power in their lives. For most students, science classes are simply another place to sit and get credits—not a place to gain a new way of thinking.

Unlike the tremendous intellectual industry that K-12 educational research commands, research on the effectiveness of higher education and its content pales by comparison. It is worth remembering that college attendance of all sorts may be commonplace, but degree recipients stand at only about one in four in the USA. Resistance to evolution is not really created at universities—it tends to come there, and in stark examples I will share it is sometimes fostered there. Higher education is but the final piece in the capacity of the cultural system I am assessing. But for its part, there are some big problems I have seen that, if typical, need work. If I have done any service to the proposition, it will serve as a solid example of illustrating problems by "creating a continuum of education" from kindergarten to the university (Jensen 1999), both in practice and in research agendas.

In interviews, I worked back through student life histories with science, religion, and how they came to their positions regarding evolution. To further enrich these student stories, parents, past teachers, college professors, and community members shared their stories of how they have related toward evolution, and how they make sense of it now. At the same time, as a participant, I observed their classes documenting content and practices. What I have amassed are dozens of interconnected stories which speak to the practice of evolution education at a university, and build temporally and socially into a web of significance. By this, I aim at providing a fuller picture of evolution education as it takes place in its cultural context. From this, we get a clearer picture of the different meanings and associations that students make, and the degree to which evolution is a token in bigger ideological clashes. With this kind of data, the concept of *world* and the pitfalls of attempting instrumental worldview change are vividly illustrated.

2.6.2 Where "One" Goes to College

If one believes Levine (2006), and for the time being I do, there is a strong indication that the most normative picture of student attitudes toward evolution is taken at a regional college. The greatest numbers of bachelor's degrees come from this type of institution. Regional colleges usually have pasts rooted in old state normal schools, historically the training sites of the nation's teaching force. Again, reflecting on the degree to which evolution is or is not being taught in the USA, this type of institution creates the largest share of our teaching force. From this thinking also follows the most normative college biology classes, student activities, educational preparations prior to college, religious affiliations, and family dynamics—all key to my story. There is no ideal "Middletown" (Lynd and Lynd 1929) for college campuses, although I feel this comes as close as any you might conjure.

Somewhere in the Ohio River valley sits Mason-Dixon State University. Mason-Dixon does not quite know where it is. Depending on who you ask, it is in the American South, Midwest, or North. Poor Mason-Dixon—from the onset, it has identity issues!

Mason-Dixon State University served well as a field site. It is a regional university—closest to a "norm" of college experience for the largest number of college attendees and generates the largest number of future teachers. Although not usually the marketing focus of prestigious universities, colleges of education are the gatekeepers to most aspiring teacher's careers. Whereas larger research universities and smaller liberal arts colleges may also produce teachers, these campuses are the cultural home of teacher education in the USA. Mason-Dixion State and all students, teachers, college faculty, and community names are pseudonyms.

Sociological matters such as demographics and desirability mattered. Mason-Dixon would be a "goldilocks" institution with neither too much academic prestige nor complete non-selectivity. It is attractive enough to be "okay" to a broad swath of student socioeconomic backgrounds without attracting too many from the extreme fringes. Accessible to both rural and urban, Mason-Dixon lies at the periphery of a regional metropolitan area while also being within an easy drive for the very rural population it serves just past its campus. The Answers in Genesis Creation Museum recently opened in the region, providing an institutional presence toward potential campus conflict by Creationist advocates on campus. Answers in Genesis now commands the largest and most influential Young Earth Creationism organization in the USA, and possibly the world.

What is clear, as I will discuss, is that effective or ineffective evolution education—*by this a student's disposition toward the concept*—is largely set prior to coming to college and changes little during the process. For many students, there are conceptual blocks in the way, placed by the contingency of their lives. For the students who shared with me the compelling stories that follow, I wish them well on their journey. As a matter of reflexive honesty, it was, at times, very difficult to not engage fully in an open discussion with these students regarding evolution. Evolution is, for me, one of the most profoundly elegant yet simple ideas. As a teacher, I wish to share with students this beautiful insight into the natural world. But first, let us attend to the world of people.

2.6.3   *What We Talked About*

Naturally, I would be curious about how ideology affected receptivity toward evolution. As educational anthropologists such as Varenne (2007) and educational historians such as Kliebard (1987) have reminded us, some concepts within education draw on deeply held knowledge from outside education. Education "writ large" or as the "transmission and receipt of cultural information" (Spindler 1987) naturally is embodied in the ideas and practices of individuals with commitments largely outside the schoolhouse, especially in the case of evolution. Given the limited ability of American public education in practice to discuss religious matters, let alone the subtleties of comparative claims of ontological stances, what ideological battles take place within the college classroom, where do competing ideologies come from, and what changes take place for students during the process of college education? I was both keen to observe and discuss with students how they negotiated this process.

Discussing students' educational histories, I was interested to apprehend to what degree dispositions toward evolution culturally reproduced, how often and to what degree this changed, and what other phenomenological events accompany a change of worldview? What cultural costs and benefits were associated for them when it came to accepting or rejecting evolution?

The comprehension of evolution requires for most a distinct set of foundational concepts (old earth, abiogenesis, and mutability of species), much of which exists outside of the worldview of many people. Would the process of an effective education in evolution open the door to a reconceptualized world, or tax some people with existential anxiety? What does this process "look" like phenomenologically? What was gained and lost along the way, and for whom?

From the phenomenological illustration I draw regarding how one learns about evolution emerged ethical questions about education practice and social justice. What is the most socially just position from which to approach evolution education, given the seemingly unavoidable conflict with many people's theological commitments? When the majority of Americans have already made at least a passive commitment against evolution, from what stance does the authoritative, but numerically small, corps of academia proceed with evolution education? The answers to these questions animate the body of this book.

2.7   Organization of the Text

Academic discussion of the public relationship toward and broad-based rejection of evolution, in my assessment, is limited by the philosophical positions that underpin the dominant voices of science. One issue at stake for me, as I have begun to unpack in this introductory chapter, is that the socialization of scientists and science educators rarely includes an education and understanding of why most people could simply care less about their work. Demonstrative of the ridiculous lengths some scientists have gone to advertise the overarching value of science for the lay public, they miss the larger domain of culture in which people operate. Reviewing the research that has already been done concerning student, teacher, and community disposition toward evolution, there is clearly a lacking commitment toward presenting evolution education unqualified by religious deference.

Turning to the view that education is best described as a cultural process, my methodological rationale of how a view through the entire educational structure gives us a better grasp on how and why evolution education could come to be so stunted. Starting with the university as a good place to apprehend people's under-standing and relationship toward evolution, I discuss the relationship between regional master's degree granting institutions, and their prime role in generating the majority of teachers in the USA.

In Chap. 3, I will begin by illustrating a few individuals' strong resistance toward evolution, and describe the cultural basis of this. Following Flyvbjerg (2006) program of *phronetic* social science, I will examine what an extreme case can teach us about the general. To do this, I will start with a student religious leader who heads a congregation of hundreds of students who actively stand against evolution within the content of their campus faith programming. It is important to begin this way as I will detail, because Creationists have had their epistemology socially shaped against evolution, but are the least likely to recognize the sociality of these influ-ences on their lives. I will next illustrate that the sociality of resistance to evolution shares a present but undiscussed phenomenon.

Creationists experience existential anxiety when challenged to move past the absolutism of their epistemology during evolution education. The interviews I conducted first unpacked student conceptions of evolution and what deficiencies they saw in the theory, and then I asked them to consider the effect of evolution "being true" on their lives. Considering existential anxiety requires analytical tools of a much broader repertoire of philosophy than that which a philosopher of science or science pedagogue usually considers germane. Again, in this way, rejection of evolution is not a "scientific" problem. These examples will work to wean readers past the cognitivist "conceptual change theory" models which as I assert miss quite a bit of what is going as students resist evolution. That is again, the phenomenon, while experienced individually, is produced socially. But as we know, people's lives and understandings are not fixed, or quite simply education would not work.

So how are these Creationist students and their unique philosophical conundrum distinctive amidst their peers? As I will continue in Chap. 4, there are essentially three meaningful positions toward evolution one might encounter when speaking with Americans. Using sociologist of religion Robert Wuthnow's (2005) ontological categories of religious exclusivists, spiritual shoppers, and religious inclusivists, we can more clearly see how arguments against evolution come from ideologically "typical" positions. I will articulate a framework with which we will think about religious ontology vis-à-vis evolution. With this, we can begin to outline what it will take to move evolution education forward. In my adaptation of Wuthnow's work, I add a secularist category and collapse his spiritual shoppers and inclusivists into a position he more recently (2009) describes as "Both/and"—people who adaptively use religion and science in their lives in often contradictory but socially ecumenical ways.

In Chap. 5, I rejoin our ethnographic discussion respective of the recent move in science educational research to consider worldview. As we know, as life changes, ideology and worldview can change. Receptivity toward evolution can follow suit. To illustrate the stakes and terms of such change, I present three case studies that show how ideology and commitment toward evolution change when one's worldview does change. That is, moving one's position toward evolution may require quite dramatic ontological shifts. These are things neither easily done and almost certainly not quickly, nor instrumentally "taught." In fact, it makes better phenomenological sense to think and talk of how evolution *can* become reasonable to a person. In the careful retelling of these three case studies, the often painful contingency of student lives shows how this does happen. Using Anne Swidler's (1986) sociological model of "strategies of action," I use these case studies to detail how the structured possibilities of our cultural lives to a great degree determine the parameters of possibility we often see for ourselves.

In Chap. 6, we back the focus of our gaze out to the campus environment as a whole. We will meet faculty in other departments for whom evolution (both for and against) is a rallying point by which the politics of evolution education on campus are played out. We will meet the Mason-Dixon biology faculty and unpack how their own experiences do or do not make for a productive dialectic for evolution. I will then extend the discussion to the political tensions at play and the rationale for Mason-Dixon to hold a "Darwin Day" outreach program for local schools. In the end, we will have spoken at length about the influence of both science and religion on students and campus life.

Following this, Chap. 7 details a thematic analysis of the social perspectives and educational histories of students generally toward evolution, working us toward a richer description of what the normative case of evolution education usually looks like, along with the odd extremes. From this, I connect the discourse of campus outreach to the politics within schools—and share the views about evolution held by Mason-Dixon students' past science teachers. In some of these cases, educational practices are likely in violation of standing law.

Chapter 8 draws together and closes my discussion by considering the two forms of *Truth* by which the American scientific and religious discourses appear to be solidly rooted. By introducing a third perspective via a mythic scene from Americana, I consider ways to open dialectical spaces for growth and possibility regarding evolution education. In doing so, I open discourses that move evolution education forward by considering what ungrounding some of the deepest American cultural attachments might mean for science, religious practice, and education.

Chapter 3
Evolution and the End of a World

You may run home for a long time
Run home for a long time
You may run home for a long time
Lemme tell you, God Almighty gonna cut you down

You may throw a rock, hide your hand
Working in the dark against your fellow man
Sure as God made the day and the night
What you do in the dark will be brought to light

God's Gonna Cut You Down (excerpts)
American traditional

3.1 The Phenomenon: Resistance to Evolution

This chapter serves two purposes. First, I put forth examples of Creationist students experiencing existential anxiety in the face of evolution education. As I show, this provides a more nuanced explanatory framework than much of what science educators have been writing and researching regarding Creationists in education. What I am after is a more robust description of how Creationists experience forms of educational knowledge that are both outside their day-to-day vocabulary and that ostensibly works to dissolve and reform parts of how they conceptualize, at a base ontological level, their view of their world. What rationale, more deeply situated in sociocultural theory, might we offer that illustrates a clearer picture of the *phenomenon* of rejecting evolution?

I start with an example of a Campus ministry student leader's experience and illustrate how discourses usually rejected from classrooms are quite useful in interpreting Creationist students' understanding of evolutionary science. I show how I came to this line of inquiry through preliminary research, and then show the thematic chorus in which Creationist students spoke regarding this anxiety. Secondly, I ground these students' experiences within the philosophical framework of existentialism that addresses such issues.

D.E. Long, *Evolution and Religion in American Education: An Ethnography*,
Cultural Studies of Science Education 4, DOI 10.1007/978-94-007-1808-1_3,
© Springer Science+Business Media B.V. 2011

3.2 The Last Outpost

It is about 10 a.m. I sit in my parked truck for a few minutes waiting. Looking around, here and there other students sit in their cars doing the same—catching one small moment of solitude in the abrupt transitional space between work and school, or home and school. For some, this is the extent of their campus extracurricular life. I had pulled into the Campus Christian Center's parking lot to make a quick check as to the origins and intent of a flyer I had picked up while walking around the Natural Sciences Building. On it, curious questions were raised, as presented in Fig. 3.1.

Certainly, this being a fairly large public college, such an outreach effort intended to bring students in for fellowship and discussion that clarified any tensions. Pulling the door open, I spied what struck me as a heavily used recreation room, spotted with ping-pong tables, public college dormitory-grade sectional sofas, and an oversized television. With no immediate signs of life, I turned toward a side corridor that held three cramped offices—one containing a few students. "Hi, I'm trying to find someone who can talk with me about the history and role of this organization on campus and about your advertising in the sciences building…." Looking a bit dazed by my direct approach, one of the students hanging out smiled and confirmed that in the next few days she would have time to talk to me. We exchanged contact information. As I glanced about taking in the scene, two books stood out to me. Sitting tidily atop an otherwise messy pile of papers, I saw two copies of *Evolution Exposed*. Later, a quick bibliographic search lists this as an *Answers in Genesis* publication (Patterson 2006). After a week of no response, I returned—this time armed with a glance or two at the organization's web site and historical presence in the campus newspaper.

Once again pulling the doors open, I spied three heads that popped up from behind the sofas, apparently slunk down and reading. I quickly turned toward the offices and saw an open door and a cluster of people. "Hi there—I'm looking for a Matt Leslie, the student leader of the Campus Ministry." With no words, two of the three students, both women, backed up and a tallish eager young man whisked his way forward showing me to his office. "I'm Matt—what can I do for you?"

We sat down uncomfortably but politely. People doing "research" likely do not come to these parts much. There was at once both a look of suspicion and hesitation on his face along with an awkward eagerness to speak plain *Truth*. I proceeded to explain my presence. After getting past the preliminaries and sensing a great easing of tension, we began to chat about the Center, its role for students on campus, and their vision of outreach. "We've been on this campus as long as it's been around," he explained. Continuing about the intent of the campus ministry, "college students don't like to go to church, as a rule kind of—so a while ago a bunch of churches decided to go to them, to actually move the church to the campus." As he detailed, the campus centers are a sort of last outpost before students enter (are lost to) the "world". Asked about how they differed from other campus faith organizations, he clarified that "we're the only organization on campus funded by outside churches," and as he put it:

> We're here to share the gospel with anyone in the campus community—students, international students, faculty…in reality, we've got atheists, Buddhists, and Muslims who all hang out here—it's maybe only like fifty percent who are Christians—we're just a bunch of college students who like to hang out together.

Fig. 3.1 Campus Christian Center's outreach flyer in the Natural Sciences Building

I was a bit skeptical of the matter-of-fact "everydayness" of his explanation. Pushing the issue, I asked for his clarification as to the theological differences between the Campus Christians and the other Christian groups on campus. As he explained, their focus was on outreach. "It is our ultimate purpose, as I believe the scripture says, to love God first and love people second…that's what sets us apart." As he made immediately clear, unless his organization was quite unusual, evolution was going to be problematic. "We take the Bible, every word of it, very literally, and that has to do with the Creation story. There are some Christian groups in the country today that have kind of fallen away from that and are going some other ways." He went on to contextualize how often they hold events, who comes to them, and the types of programming they focus on. Out of this the following stood out:

> For us, evolution is huge. Anything that draws people away from our faith, we're interested in discussing. We are trying to bridge gaps. Two weeks ago we had a speaker who talked about what we could do as Christians to care for—without the conflict—people who have other beliefs than us. He had a guy who was an atheist—or used to be an atheist—that he brought with him and interviewed him. He explained that in everything that we're doing, we're trying to get away from that whole conflict idea. There's been so much of that in the past and that's kind of what we're viewed as.

I next turned to questions about the group advertisements for which I came in to talk. I asked about what was meant by the world being "messed up," as described on the flyer. Matt explains:

> Sin. Our purpose is in ultimately sharing the gospel, to understand why we need God, why we need to believe in Christ. First we have to believe that we are sinful, and not worthy to be in the presence of God. We can't get to heaven because we're all sinful—Adam and Eve?—that's what's messed up about the world.

When I asked him about the science and Bible in conflict question: "Our point is that there shouldn't be a conflict. That was the whole point, we did a set of weekly readings…it went along with this whole packet that asked those questions…." As I would later discover, the packaged materials were part of a canned evangelical teaching packet from the Central Christian Church in Las Vegas. Upon my own web site searches, it was clear that the Campus Christians had followed the scripted plan down to the copied *WHY> ?* typeset (Fig. 3.1). Having gained a comfortable place in the interview, I turned to the business at hand.

3.3   On the Brink of a "Destruction" of World

Well before meeting Matt, in the early stages of my research, some issues began to stand out that illustrated a deeper complexity to the phenomenon of resistance to evolution, which previous studies have said little, or even considered. In one of the more compelling early interviews, a nontraditional student named Esther reacted to the following lead-off question: "Can you describe to me your relationship to the concept of evolution in your life?"

> My parents were missionaries, and I've grown up Christian my whole life, so like, we didn't have that debate in school like, '*should* we teach evolution'? I personally don't believe in evolution, so it was kind of a shock to me when I had just come to college. I took intro bio in my freshman year…and I went through that whole 'what do I believe…is it my parents or is it other things'…I think that was my first contact with it. So, [evolution] was like huge in class, and he presented it as if there could be no other explanation…and, I mean I remember him saying that if you believe in God creating the earth, then pretty much you're an idiot. And he obviously didn't use those words, but that's pretty much what he said. And I remember thinking… *What is he talking about?*…like, he should just go around and ask people to raise their hands 'who believes in this and who doesn't'! …and then I was thinking about… *are we going to do this for the theory of genetics?*…that's presented in pretty much exactly the same format.

Was this internal dialogue, fraught with anxiety and unspoken conflict, indicative of something broader that I might investigate? As Esther continued, it was also clear that traditional measures of educational success such as grades were not appropriate for estimating receptivity to evolution:

> I take those really big classes, because it's really easy to excel in those huge classes. I mean, I got like a hundred on every test. You have to be an idiot pretty much not to. If you just sit, and you listen to what they're saying, and you know how to take tests, it's very easy to do well in those classes.

With Esther's internal dialogue before me, would it be possible to more fully explore this thought space? For her, an atypical student by way of her life course, the experience of evolution education in the college classroom sounded like one of alienation or marginalization at the hands of a professor's assertive stance in the class. The problem with this view is that the liberal academic tradition does not usually consider itself as one that marginalizes. As a "church of reason," its duty is to expand minds, not set them fleeing. I set to devise a course of interview questions that would prompt students to explore this relationship. We would engage an individual's conception of evolution in ways that our social norms of dialogue regarding evolution usually do not prompt. Rather than avoid evolution, I would go right at it, untying what speaking about it meant in the lives of those who reject it out of hand.

Back at the Campus Christian Center, I explained to Matt the kind of research that I was doing and that his views regarding evolution could be important for moving the conflict metaphor of science and religion which he had addressed (and with which I was concerned), toward a more productive dialogue. I asked him to walk me through his early experience with evolution in school. As he explained, when evolution was discussed it was not presented as "This is the way!". Rather, as he explains, mimicking his teacher: "This is the way that science sees it; this is the way that the church sees it." Leaving room for interpretation, his teacher seemed to be able to negotiate competing ideologies in class, footnoting the reference to "the church" in my mind. As Matt took from this "We had to be able to say what science said, but I don't know if [the teacher] personally believed that, so we kind of left things open." Matt went on to explain what seemed to be an unexceptional biology class, but which did cover evolution for at least 2 weeks.

I turned to whether evolution was ever discussed at home or not. Matt responds quickly as if sensing an opportunity:

> Oh yeah. Not as far as learning about it…it was more about talking about how stupid it was because my parents are both *strong* Christians…so I didn't learn a lot about evolution at home, *definitely*. Most of my knowledge of evolution has come since I've been in college in personal study and sermons at church.

As Matt explained to me, his 2 weeks of high school evolution education were topped by about a month and a half while in college. Seeing evolution as something that comes in from the world outside the family, he talked about how evolution was treated as a subject. "I brought it up several times just asking questions and stuff… but it's never been discussed in my family as like—an option…we never sought out to see if it's like, real or not." Matt could remember at least three times when evolution was the subject of the sermon at church. "In my 22 years of life I've probably heard at least three sermons specifically about that on a Sunday morning… It was probably more about 'teasing' it."

Discussing the sociality of his life, and how evolution has come up with friends in his social network, "a lot of my friends believe in evolution, even some of my Christian friends believe that—in theistic evolution—the idea that God caused evolution to happen." This influence has had an effect. "I've even leaned towards that in part of my life." As he went on, it became clear that evolution is a recurring thorn, especially given his leadership role in the Campus Christian Center:

> We often discuss that because I think it's central to what we believe… and what we believe is central to our lives. Any time that I share my faith with someone [evolution] comes up. If I begin to talk about Christian beliefs it comes up automatically because it's a big hurdle that people have to get over for them to believe there's a God. I've read a lot of books.., well not a lot of books, but I've been to a seminar on it and we had a big thing on campus last semester, a Mock Trial about it. Evolution is just kind 'around'…it's underlying in almost everything that we do or discuss…it always comes up.

Moving on past Matt's life and the role evolution played in it, I turn toward his understanding of the concept. I asked him to detail to me what evolution is. "Macro evolution …would be the idea that…" (*trailing away*). I made note of his immediate invocation of "macroevolution" as this would be a common theme for Creationists:

> Evolution is the idea that things change over time. Macroevolution, as I see it, is the idea that we started from absolutely nothing. That there was no life on earth and the primordial ooze and whatever you call it…sparked life and things have come from that and some animals evolved and humans have evolved from apes. That's my understanding. I believe in micro-evolution…survival of the fittest, that's pretty obvious. We change…I change, I change over time…I don't think you can deny that. But as far as macroevolution I don't see that.

The distinction between microevolution and macroevolution is a commonplace contemporary emphasis made by Creationists to preclude the possible mutability of Biblical "kinds." I then asked Matt whether he finds the science behind evolution as sound. "Well it hasn't been proven…totally. Obviously I don't believe that it can be because I don't believe that it's the *Truth*."

Matt turned existential as he contemplated the evidence he sees proffered from science and his faith. "I believe there's all kinds of evidence to point to [evolution], and I won't know why that is until I get to eternity." I asked him to detail what he sees lacking in the science behind evolution, as he understands it:

> I'm trying to think…(*long pause*)…they haven't found that link…between non-life and life. It just doesn't happen without a Creator…there just has to be a Creator to put that life… (*pauses*) Just the sheer complexity of living beings, even non-living things. It's just so… everything is just so complex. The great analogy of all the parts of a watch falling into a pool of water and bouncing together somehow… I don't care over how many years - that stuff doesn't come together and make a watch that ticks….that stuff just doesn't work.

Matt immediately segued into a discussion of recent books he has read, notably Francis Collin's (2006) *Language of God*, which he found compelling, but ultimately found lacking for its far more liberal interpretation of Biblical text. In the *Language of God*, Collins puts forth a theistic evolutionary conception of the "BioLogos," with which he hopes to bring people into an understanding of both evolutionary science and the Christian God. But Matt's position seems immovable. "It is just unbelievable… there's lots of evidence that points to [evolution] now. But I hold so strongly to scripture that I can't make that jump from what scripture says to what a scientist says…."

I moved toward the crucial turn I would make with each Creationist interviewee. After having each detail of what they saw as deficiencies or improbabilities for evolution, I put forth a proposition that would rend this view asunder. Asking Matt to consider a thought experiment, I relay from the transcript:

> *What if you woke up tomorrow and everything you pointed out as lacking in the science for evolution…what if you found out that it actually had happened. That there was enough good science there to support it, and evolution had actually happened?*
> (*Long pause. I noticed his face has flushed*)

> Well…it would be a big deal. You know, Francis Collins talks about how in the church in the past, they kicked Galileo out for believing whatever his theory was about the earth revolving around the sun…and I would say, speaking for myself and all of the Christian church, it would not ultimately take away from what I believe. It would cause me to ask a lot of questions and a lot of rethinking of things. I'd probably be a little confused….but

that's what's cool about God, I never have to be afraid of any of that…or worried about any of that…it wouldn't be too big of a deal.

Having reacted to grandeur of the premise, and then invoking the plasticity of God as his way out, Matt continues to ponder the social implications of such knowledge:

It would change the way I talk to people and the way I present the gospel. If my parents couldn't see the same evidence that would change my mind on that. It would have to be so overwhelming that everyone would see it, so it probably wouldn't be as big of a deal. It would be a big deal if I changed my mind and they didn't…because believing in God as the Creator is so central in what we believe that that'd be a big deal.

Without batting an eye, or conceiving of other ways to think about evolution, religion, and *Truth*, Matt's utmost concern centered on the mutual belief system shared with his family and social life.

Given Matt's clear opinion regarding evolution, his experience in school was of interest to me. When I asked him to relay how he negotiated evolution being taught, his reception was not as clear as one might presume. "I'm probably a bad person to ask because I hated it the whole time…but for me it was very helpful. It's a big thing that I have to deal with in my job." As Matt clarified, part of their campus outreach involves bringing in more to the antievolution fold. "I have to deal with it on like a weekly basis. There's a lot people I have to deal with that believe in evolution." As we talked, although the spirit is one of fellowship and welcoming, the message and obligation are toward strict Biblical literalism. Using a recent guest speaker on evolution and Creationism as an example (and the reason that *Evolution Exposed* was sitting on his desk):

That's why that guy came and spoke a few weeks ago. We have to as Christians understand evolution so that we can fully understand what we believe. It's frustrating to me that everybody doesn't believe what I believe…but I understand it and it's something to deal with. I wish it wasn't taught because it's not the *Truth*.

Encroaching on my very next question, Matt restated his feelings regarding the status of evolution in education, and how this relates to *Truth*:

I hated it because it's not the truth. It has not been proven. Science to me is—if you're going to present something as proven by science it should actually be proven by science. What we've done is just jump on [evolution] so strongly without having the ultimate proof that we need to believe it.

As Matt continued, I continued to take note. For some, the implication of evolution is no different now than what Darwin knew of his impact upon publication. Epistemologies and ultimately ontologies grounded in a fixed view of *Truth* would find such knowledge incommensurably wrong, perhaps even evil. This became clearer as Matt continued his cathartic dialogue: "I guess there are people working full time to get there…but to teach children and anybody, really, something as the *Truth* when we don't know that it's the *Truth*—is kind of disgusting to me." The issue is clear enough for Matt. Unlike many who confuse formal theory with the colloquial usage of theory implying hunch-like assertions, Matt knew full well what evolution as theory implied:

If they want to present [evolution as theory], that's OK, but a theory in science is basically proof. I mean a theory in science has got to be pretty proven to be a theory…and to present evolution, it just doesn't fit the definition that evolution has. If they're going to present it that's fine, but present it as a viewpoint. Present it as something that does have a lot of evidence pointing to it but we just don't know. If that's the way they want to present Creation science too then I'm fine with that. I look at evolution as a faith, not as science, I think it takes more faith to believe in that as science.

Pausing for a second, it is important to remember the failing of most previous research concerning the efficacy of evolution education. Although unbeknownst to him, Matt had just opted for Nietzschean perspectivism for scientific theory all the while requiring his Biblical *Truth* to stand behind it as supreme. No other more moderate theological interpretive repertoire interceded. Kuhnian paradigms were not ready at hand.

The context of Matt's life, and the practical rationale by which he came to his understanding, is something that needed to be accounted for. This investment in his world and his understanding of being clearly steered him away from evolution. But jumping upon someone for having confused or errant notions of the paradigmatic state of science is as futile in the end as criticizing a Nobel laureate particle physicist for not being up on all the details of evolutionary ecology. I would not, and did not, engage in any on-the-spot critiques of students understanding of evolutionary theory, and as a reminder to my readers, you can go ahead and let that go yourself, that is not the issue here. The issue for Matt is not one of science—it is of meaning and ultimate purpose. Unless you make the scientistic move of equating science as the only appropriate way to think about the world at all times, science cannot give ultimate meaning or purpose. To do so is to hold science up in the same spirit and calcify it with liturgical trappings of a faith.

For Matt it is clear, and a no-brainer assertion, that scientific knowledge that strays from or contradicts his Biblical literalism then becomes invalid science. The matter here is not one of someone needing to be presented more evidence. Rather, in this mind-set or worldview if you like, there is no way evolution could ever possibly be true. Rather, what are we to make of his take on the prospect of evolution being true—it being "a big deal" and it making him "confused?" Is Matt an isolated case? Surely not, as we know. But he is indicative of something more general, an ephemeral something which has flown under science education's positivistic radar. We are, as you will see, at the precipice—the place where a world ends and possibilities begin to run out. This is an ephemeral world of our social and imagined ontological space. Aware of floating in this space, we have, like a nightmarish fear of falling, anxiety at the idea of that which gives our social world its significance dissolving away.

> As a state of mind which will satisfy these methodological requirements, the phenomenon of anxiety will be made basic for our analysis…Shrinking back in the face of what fear discloses—in the face of something threatening—is founded upon fear; and this shrinking back has the character of fleeing (Heidegger 1962 [1927], pp. 227–230).

3.4 *Flee!*

During the course of the spring 2009 semester at Mason-Dixon State, I sat down with 31 students and interviewed them regarding their life histories with the concept of evolution. Some had a strikingly affective relationship toward the concept. Except for one discrepant case, the worried tone and potential risks were only described by Creationist students. Tyson, a biology major, talked about his early exposure to evolution. "Evolution probably came up before I took my first serious biology course in high school. It came up way before that. It was always presented to me as not true." A self-described Creationist, Tyson's experiences at Mason-Dixon were broadening his perspective in ways that made him consider other people's ways of *being-in-the-world*:

> It's a big reason why, in the last year or two years, I really questioned it. My parents are Creationists, but I don't know if I believe this stuff. So I was like going through trying to figure out what I believe. They all say it's scientific, but I'm like, 'really'? A lot of the reason I'm Christian or a Creationist is because my parents are and my friends are. Social life plays so much into that.

It is important to note, both in this case and for studying American education in the current context, that Tyson was homeschooled. Homeschooling in the USA today is overwhelmingly dominated by those who "self-identify as evangelical Christians...Most homeschoolers will definitely have a sort of Creationist component to their home-school program" (Lovan 2010). Within Tyson's story, an additional detail regarding the rhetoric of Creationists and where the "debate" between evolution and Creationism might be heading surfaced. With no prompting whatsoever, Tyson deployed worldview as a conceptual schema in his discussion:

> It would take a lot of work for me to change my worldview, a lot of time, a lot of alienation from friends. That's a big part of that. I'm always looking and trying to figure out what the *Truth* is. It's hard. People tend to believe what they've chosen to believe. You believe in something, and you'll find a way to believe it. That's where the presupposition on worldviews comes through. If you have something against God, you don't want there to be God, you're going to be a lot more prone to saying evolution works. If you're saying, 'I love God. My family believes in God'. Then you can be a lot more prone to say creation works.

Tyson's speech graded on frenetic. He had a lot to say and it poured out of him. When I asked him to reflect on a specific instance of discussing evolution with family, he had a clear internal dialogue which he relayed:

> Just recently, I had several conversations with my mom about being broken myself. I said, 'This is scary. Is this true what I've been believing for years about this whole Christianity thing? Some people are Christians, but they're not Creationists. If evolution exists, can I still believe the gospel at all'? Then we talked about it. I said, 'Mom, I've been reading this argument over here'. I'll bring up things and we'll talk through them. She'll tell me what she has heard. I'm like, 'that doesn't work.' I usually play a lot of Devil's advocate. I'm trying to get to the *Truth* and not just believe what she tells me:

With Tyson discussing such a rich experience, I posed to him the same hypothetical as I posed Matt of the Campus Christians: What if evolution actually were true—what would that mean to him? Directly from the transcript:

> It would be a complete crisis. It would be really tough. I've asked people this. Some people don't believe evolution because the Bible says so. I'm like, 'You have to have some rational reason'. People believe the Qur'an and all that stuff. We don't believe that. So then it would be a big crisis. The Bible teaches evolution is not true. If the Bible is not true, then there is no afterlife. I don't know why I should be good in the first place. There's a lot of conflict there. I'd have to try to find some other evidence. If there was no evidence, I'd either give it up, or I'd be too lazy to give it up because I want to keep my friends.

And how does that make you feel?

How does it make me feel? I guess I don't feel like I really feel anything, but I probably do, but I just don't recognize it. I've thought about it so much that I'm kind of numb to it.

So would it have an effect on any other parts of your life?

Definitely. Hugely. My relationships and who I'm friends with. What would happen? It would be a time when you'd have to make a decision. Am I going to keep believing what I believe? Am I going to search for more evidence, or am I going to throw out Christianity? Based on what I do, my relationships with other people would be kind of dependent on that. If I say, 'No, I'm throwing it out the door', and they say, 'No, I'm going to keep it', that's a big barrier between relationships.

Adding to Matt's concerns but with perhaps a bit more concision in his self-reflection, Tyson added another voice to the concern of worldview. As he framed it, parts of his sense of being and purpose would be in "complete crisis." He described his own knowledge base, exemplified in discussion with his mother, as "broken." Of note now, and as I will discuss later, Tyson makes a set of critically slippery moves in his narrative. For evolution to be true, his interpretation of the Bible is likely untrue. If this is so, his ethical system breaks down—"I don't even know why I should be good." The reward of his ideal, an afterlife, dissolves. Again, no more moderate theology intervenes. Crucially important, and completely missed by almost all prior research, is the social dimension to this phenomenon. For a Creationist, learning evolution and reframing one's ontological stance, or having it reframed for you, has definite social costs. Now, as we will explore later, it may also have benefits, but when standing at this brink, one only sees the possibility of destruction from a fall or the possibility of self-annihilation by throwing one's self off into the abyss.

One like-minded student who expressed similar feelings was overtly contemptuous of what she intimated as sinister motives within the sciences. Andrea describes:

Evolution has kind of a negative connotation to me. I feel that a lot of the facts that people use to support it are not really *factually* based, and I feel that the research that is done is kind of…I don't know…not as accurate as it should be. So I'm not a real believer. I feel that it's kind of a twisted idea of what *really* happened, I guess.

And when I posed to her the possibility of evolution being true:

I think that it would… (*Andrea pauses mid response, voice audibly broken by either a swallow or having been emotionally moved*)…it would be a big shock. I think that I would be very skeptical of the irrefutability of it at first and that would certainly alter some of the ideas that I hold to be true, but it would not completely nullify them.

Finally, what that would mean for her life:

I would be shocked. I would be kind of disappointed I guess. Just the fact that what I had held to be true for so long would now have this big hole in it.

Shocked, disappointed, and *Truth* made into a doughnut. Andrea adds to the dour mood being put on evolution. Soberly, Andrea was also the sole secondary science education major in my interview pool.

Renee, another biology major, and "new Christian" as she described herself (she had just been baptized into an evangelical congregation) added to the dark tone these conversations were taking. When asked about her relationship to evolution in her life, she brought in another of the concepts which I laid out in Chap. 2 as crucial foundational context to understanding evolutionary theory—an ancient earth, and the associated theories of the origins of the universe:

> I've always been kind of aware of the whole origin…origin event, and the natural descent. It never seemed to quite make sense, but you can definitely see how species can—I call it adapt rather that evolve—adapt to their environment and change over a period of time. It's the whole Big Bang Theory that doesn't quite seem to make enough sense to me.

Turning toward the existential:

> It's easy for people to think things just happen over a course of millions and millions of years. But it gives people no hope in the end. If we evolved from molecules and now we grow, grow, grow, and die. What is after that, just nothing? There's no concept of a soul and that just doesn't seem to make sense to me. There's got to be something more. We need to answer for the things we do over the course of our life.

Adding to this, even more students got right to this point. Skeptical at my initial assertion that evolution being true was even hypothetically conceivable, Hannah contended: "I don't know. I'd have to see the irrefutable evidence, I guess". Continuing on by posing it as a possibility that she might consider, she was blunt. "It scares me." Why would such thoughts prompt fright? "Because my whole belief system would not exist anymore, I guess…(*incredulously, with frustration*) …I don't know." I probed forward as to how this would affect her belief system. "Well, if something came out that says we evolved from monkeys then the whole idea of Creationism from God wouldn't really work anymore."

Yet another student recollected early dissonance with the idea. Articulating the same thematic avoidance of evolution, Mary explains her understanding of the concept: "It's kind of a conflicted thing. It means a scientific way of how earth evolved, but it conflicts with my Christian views. I've never really gotten into evolution because it conflicts with my religious beliefs." She went on to articulate a crucial distinction, especially given her plans to teach.

> I feel it can coincide with Creationism, but I don't believe the entire—I don't believe I evolved from a monkey. I guess a big part of it is that I don't believe in evolution, and that's always turned me off science.

I asked Mary if she could point to what specifically turns her off science. "I think how much it can conflict with Christian views." When asked whether evolution had ever been discussed in her family, Mary explained: "When I'd come home from school and say, 'We're studying this'…my family, my brother and I are very religious. My parents taught us to be that way. It kind of got into that we didn't really believe in evolution." Mary made a distinction common to contemporary Creationists. She rejected Darwinian macroevolution while muddling the conceptual issue by supporting microevolution. "Although it can be proven—it's not necessarily evolution I don't believe in. It's Darwin's version of it. That's kind of how it came up at home discussing evolution versus Creationism."

As with other Creationists, I proposed to Mary to consider the implication of evolution having happened: "I think it would really alter my religious beliefs." I asked her in what way her beliefs would be altered:

> The way I was taught of evolution that it comes from one cell is true, then it would dispute the theory that God created the earth, humans, plants, and animals. If everybody came from one organism, then God didn't create everything at different times at all.

In the face of such a prospect, I asked Mary to consider how that might make her feel, imagining the effect on her life:

> Confused. It would change a lot of my beliefs. It'd be something I'd have to think more about. I'd have to think more about what I'd been taught in church. It would definitely affect my spiritual relationship. It would change the way I would have interpreted everything I've done or believed the last 27 years.

The arrangement of meaning toward evolution was taking on a clear theme. Creationist students were reacting to evolution as though it directly undermined or invalidated the whole of their world. Whether the scientist or the science pedagogue disagreed with the reasonableness of this position was quite beside the point. The movement of this or that piece of scientific knowledge was not the issue as those moves clearly did not matter. They could be rejected piece by piece. On first glance, this is not really news. Teachers, professors, and some research strains have reported this for decades. But evolution, no doubt part of the narrative of home and church, was to be avoided at all costs. But why?

Even though I had not included it in my questions, students began to link evolutionary theory with the origins and end course of the universe. Responses such as "no hope in the end" or equating ontic death with "nothing" began to take on a clearly existential tone. Thoughts like these "scared" the students. They feared that their "belief system would not exist". The "whole idea" of their faith system "wouldn't work anymore." Naturally, and almost as a comical relief, the flatness of being "turned off science" was reasonable given the tension these students were relaying. No wonder they felt "confused."

This specific type of relation to faith and science is not without philosophical precedent. Philosophers working in the existential tradition from Kierkegaard to Heidegger to Sartre have referred to the type of feelings these students express as existential anxiety, the fear of confronting nothingness. This turning away from nothingness, a fleeing from confronting certain ungrounded or "abyssmal" concepts, is natural due to its disorienting character. Tillich (1952) describes this generally as the feeling one experiences upon becoming aware of the possibility of nonbeing. Caught between the reality of the physical world that science works to articulate and the phenomenal reality that some people describe as directly experiencing a divine presence in their being, a crisis ensues.

Tillich also describes the negative form that this relationship can take, notably influenced by Nietzsche. For Nietzsche, this insistence on the absolute, an insistence on *Truth* in the face of science's encroaching universalizing acid results in a form of nihilism. This particularly strong form of nihilism holds the course steady toward a singular *Truth* against the mounting evidence to the contrary. As the dissonance of adjudicating competing spheres of understanding amplify for some, it may "drive the person toward the creation of certitude in systems of meaning which are supported by tradition and authority" (Tillich 1952, p. 76). Creationist liturgy has tradition and especially authority in their fundamentalist churches in spades. But as Tillich knew, "undoubted certitude is not built on the rock of reality." He saw this type of certitude as a pathological turning away, a turn from the influence which science has wrought on the whole of the Western tradition. As is worth quoting at length, for those unable to move with science's influence, Tillich warns:

> There is a moment in which the self-affirmation of the average man becomes neurotic: when changes to the reality to which he is adjusted threaten the fragmentary courage with which he has mastered the accustomed objects of fear. If this happens—and it often happens in critical periods of history—the self-affirmation becomes pathological. The dangers connected with the change, the unknown character of the things to come, the darkness of the future make the average man a fanatical defender of the established order. He defends it as compulsively as the neurotic defends the castle of his imaginary world. He loses his comparative openness to reality; he experiences an unknown depth of anxiety. But if he is not able to take this anxiety into his self-affirmation, his anxiety turns to neurosis (pp. 69–70).

The turn from evolution and sound science is becoming more reasonable, given the world from which these students' views came. As Darwin knew and grappled with during the interim period in which he held back publication of the *Origin of Species*, the effect of his work would likely see a reconfiguring of certain concepts and commitments (Biblical literalism) within the Christian world. Amidst the flabbergasted incredulity that contemporary scientists feel toward their Creationist counterparts, Tillich continues with this explanation:

> This is the explanation of the mass neuroses that appear at the end of an era...In such periods, existential anxiety is mixed with neurotic anxiety to such a degree that historians and analysts are unable to draw the boundary lines sharply. When, for example does the anxiety of condemnation which underlies asceticism become pathological? (p. 70)

Although the more scientistically attuned toward evolution in our society are loath to see it, the "boundary lines" that Tillich describes are in fact a blur through society. Latour (1993) writes about the sociology of this as the "multiplication of hybrids"—individuals within social networks variably committed to the purely natural world and the multiplicities of the social. With the commitments we make toward religious tolerance in the American democracy, there is simply no democratic dictum that requires a person to move toward the more courageous stand that Tillich implores. Rather, the phenomenon of retreating from evolution for some is more accurately seen for the precarious vertigo of the position. For some, Darwin's *apokálypsis* was not an opening to a new world of understanding, it was the void.

3.5   Confronting the Precipice

As Kierkegaard (1957) saw, the ungrounded feeling one experiences at the edge of their ontology is one of confronting a precipice:

> Anxiety may be compared with dizziness. He whose eye happens to look down into the yawning abyss becomes dizzy. But what is the reason for this? It is just as much in his own eye as in the abyss, for suppose he had not looked down (p. 61).

With the abyss in front of you, one is variably torn between the fear of being pushed off or slipping, all the while inescapably drawn to the idea of throwing one's self off:

> Hence anxiety is the dizziness of freedom, which emerges when the spirit wants to posit the synthesis and freedom looks down into its own possibility, laying hold of finiteness to support itself. Freedom succumbs in this dizziness. Further than this, psychology cannot and will not go. In that very moment, everything is changed, and freedom, when it again rises, sees that it is guilty. Between these two moments lies the leap, which no science has explained and no science can explain. He who becomes guilty in anxiety becomes as ambiguously guilty as it is possible to become (p. 61).

Rather than confront the basis of this feeling and seeing where it goes while experiencing it as a phenomenon, the normal move is for one to retract to the safety of the known and grounded. As Matt described, "that's what's cool about God, I never have to be afraid of any of that…or worried about any of that".

Heidegger (1962 [1927]) makes no small issue of such accounts as he ascribes the pejorative *inauthentic* label to those so inclined to not take account of their full *being-in-the-word*. *Dasein* for Heidegger is the "being that takes a stand on its being." The *inauthentic dasein*, sensing the anxiety of losing the ground of the world for which they take a stand, flees to the banal comfort of what "one" (*das Man*) does. One does not authentically take account of the competing and contradictory tales from science and faith—one rejects the intrusion out of hand. The preferable stance for Heidegger, the *authentic*, is a being that takes a stand on its being with the ability to deal with the eventual changes in the cultural world that are unavoidably to come. Both Dreyfus (2005) and White and Ralkowski (2005) explore this as the individual that takes a stand in the face of "world-collapse." In familiar terms, this world-collapse can be thought of as a cultural "death" if you will. Rather than describing the physical demise of a person, Heidegger describes this form of death as *dasein* having an ontological *being-toward-death*, a general anxiety characteristic of our being that through the flow of temporal succession experiences change in a world, and takes a stand amidst it.

As Lear (2006) artfully details in his *Radical Hope: Ethics in the Face of Cultural Devastation*, there are historical examples of just such ways to make sense of one's world after the devastation of cultural annihilation. In an analysis of oral histories at the turn of the nineteenth to twentieth centuries, he considers the last few remnant Crow Indians experiencing the existential anxiety of world-collapse. For these people, to move past the devastating losses of the artifacts, land, and ecology for which their lives had intelligibility required drawing on some of the best characteristics of the prior world to give, in quite an evolutionary way, a ground of being in the new.

In the Heideggerian terminology from which Lear works, the referential totality of Crow equipment and Crow *being-in-the-world* had been destroyed. When the buffalo were gone, for the Crow, the world not only ended, but nothing else could possibly happen. No robust sense of "Crowness in the world" remained. As Lear asks of this kind of "abysmal reasoning," how is one to consider appropriate steps past the destruction of their world? Our only hope is "to establish what we might legitimately hope at a time when the sense of purpose and meaning that has been bequeathed to us by our culture has collapsed" (p. 104). So what is the appropriate way then to *be* a Creationist in the face of the reality of evolution?

For our students then, those resistant to evolution take this stand for there is existentially no other possible option, nothing else to consider or rational path to take. For a Creationist student embodying the liturgy of their life, accepting evolution as *Truth* flies in the face of the positive, commonsensical, and affirmed "practical reason" (Bourdieu 1998) of their lives. For those liberal theists and secularists who have not experienced this, an equivalent vulgar description of the emotional impact of this disorientation might be something akin to me earnestly trying to convince you that your mother or father is, in fact, a whore. In a worldview for which evolution and its requisite supporting structure are required (vast expanses of geologic time, the non-teleological changes in the morphology of organisms, and the possibility of chemical origins to life), the move toward this authentic position is too jarring, too dissonant to the comfortable narrative of home and hearth. This is not to say that an individual never makes this move—it likely happens here and there all the time. In fact, education can and should be the means by which this gets done. But for now, for our Creationists, this is the phenomenon—the anxiety at a feared erasure of their primary and utmost understanding of the world.

Julie's experience presented me the most pronounced case. A pre-dentistry biology major, she seemed to be right there on the precipice, reflecting on the weightiest of ontological issues. While reflecting on what I later confirmed as an exemplary coverage of evolution in her education before college, Julie explains that evolution was securely entrenched throughout her education, from elementary school through high school. The religious education classes at this Catholic school also discussed it. "They'd tell it to us how it was in the textbook, and then they'd try to incorporate it into religion. They presented it like, the Big Bang theory….and….a ball of dust and *bam!*…and all these planets". As these discussions interpreted for Julie, "they'd say—well, you can think of it like that and then you can say well, God created this…and evolution happened from there." Like Tyson's prosperity to speak (and quickly), Julie was similarly effervescent. I asked Julie to detail whether this process in her schooling prompted controversy. "I know evolution is like the biggest controversial topic. In psychology I was taught critical thinking so, in everything I learn, I'm like…it could be this or it could be that, so I'm always thinking….I just know that evolution is huge."

Wanting to get Julie's sense of evolution being controversial or not, I asked her to explain how and why she sees evolution as controversial:

> Oh my goodness…you think of all these dinosaurs, and everything living here, and everything being under water at one point—that's so hard to believe. But you can go outside and pick up a rock and look at all those fossils…and that's kind of evidence. But….it's

what you don't know…and even like—with death. It's what you don't know. It's so controversial…I mean there's evidence but it's like…this 'big ball of dust' doesn't *get* me, so… (*trailing off*)

There are times in interviews when you are caught not only off guard, but become transfixed by the lucidity of someone's experience. Julie had just linked her thinking about evolution to universal origins and to death! With such lucid examples, I asked Julie to articulate her association of the "big ball of dust," evolution, and death. With an excited confiding identification:

Oh my gosh,…it does!….whenever I think of like the solar system or anything like that… it's such a big fear of mine. Whenever I think of evolution, whenever I think of the solar system…anything to do with like earth or planets and everything…I always think of death… (*I interject*)

Why…why death?

(*Exhaling with self-cautious laugh*)

…I don't know!…it's so weird!…because I keep thinking I…. (*pausing, then resolute tone.*) I know where this originates. I think of a small thing—like the planet. Then I start thinking of big concepts like…the solar system…and then I start thinking of like the Milky Way and all that stuff, and then it starts getting bigger and bigger and bigger…and then like (*pausing, then speaking faster without pause*) …you don't even know what's out there… and it's so big, and like, they don't know where heaven is and stuff like that…and so, you know, I start thinking of big concepts…so I guess big concepts go with my concepts too… so heaven is like a big concept for me, like…I don't even understand it….like…how does time work in heaven, how does your appearance work in heaven…is it like, in another solar system?….or is it like… (*pause*) …you know? So, that's how I like tie that into religion… and to death.

As Julie continued, such discussions usually emerged with her boyfriend. "Over the phone I'd talk about this with my boyfriend. We'd always talk about that because…(*long pause*) …that's a hard topic because that's really scary, you know, I think the idea of…even death or anything like that…it's scary." I asked her to explain the scariness. "I don't like the idea of when I die I'm just going in the ground… it's scary to think that life just goes on and I'm done. With evolution I think it's scary because there has to be something more." Continuing on Julie's narrative came to an end. "Well…." (*long pause…hesitant exhalation*)

I have a fear of dying…and so…when I think of death, and everything….I just want there to be something more…you know, I don't want to just die…and life just keep going on… and the earth just keep going on without me…I want to go somewhere else, so… (*Julie trails off*).

If one is to take Julie's concern seriously, and I do, it is clear that the affective relationship that some students have toward evolution and other weighty concepts from science prompts a form of existential anxiety. Kierkegaard and Tillich's descriptions of this dread accord plainly with these data. So what then to do with evolution education? Part of the problem for a Creationist student is that evolution invalidates *a priori* footing of their worldviews. When experiencing evolution in education, practical theological guidance does not appear simply because scientific concepts brush up against the existential. The reformation of one's errant "misconceptions" does not take place in a value-neutral vacuum. As Mary stated, "It would change the way I have interpreted everything I've done or believed the last 27 years."

Evolution education, in its phenomenal impact to Creationists, is not value free. The emotional stakes that some people have invested in forms of cultural life inimical toward evolution certainly need dealt with more effectively if AAAS science literacy goals are even vaguely possible. Crucially, Mary spoke not only of her epistemology, but also her prior practices being invalidated—"change everything...I've done." Old-time epistemologists, logicians, and most analytic philosophers (and the positivism these spurred in the sciences) could offer nothing for Mary about what to do with her affective relationship toward her practices. Every positive memory Mary has regarding the antievolutionary liturgy of her church education is summarily negated. Merleau-Ponty's (1962) *Phenomenology of Perception* described this as the bodily portion of our being. Like the clarity of his phenomenological descriptions of the "phantom limb," evolution threatens Creationists with the lopping off of an affectively charged conceptual world of meaningful social practices. "We are imprisoned in the categories of the objective world, in which there is no middle term between presence and absence" (p. 93). What is Jessica to do with all the shared understanding and practices that would be invalidated? Does Mary simply choose the "worldly" schooling of her professors and take this knowledge home to grandma? Continuing from Merleau-Ponty:

> We do not understand the absence or death of a friend until the time comes when we expect a reply from him and we come to realize that we shall never again receive one; so at first we avoid asking in order not to have to notice this silence; we turn aside from those areas of our life in which we might meet this nothingness, but this very fact necessitates this (p. 93).

It is also clear at this point that for work-a-day scientists, the equivalent meaning they derive from their affective relationship toward science is equally fraught with concern. In other work concerning scientists protesting the opening of the Creation Museum, I have described this (Long 2010a). The implication here addresses the general attitude by many college faculties toward fundamentalists not "getting" evolution or other such perceived contentious issues while in college. Overwhelmingly, it is usually less than sympathetic. Ecklund's (2010) work on the knowledge of elite scientists in American research universities bears this out. Teachers though, by and large, do not attend such universities and as Kimball (2009) has shown, are more likely than other majors to be religious. For my interests in advancing evolution education, this status quo just simply will not do.

In this chapter, I have illustrated a deeper picture of what is going on in the minds and practices of Creationists as they encounter evolutionary theory. Far past the limits of pure reason or simple *Truth*, the fears and anxieties at play in evolution education involve the interactions of differing social fields, ideologies, and power over whose voice gets heard when evolution is taught. Although almost never part of broad discourse, evolution education prompts existential anxiety within some students. Resistance to evolution is then resistance to the conceptual framing of a different world. Working Creationists past seeing a void at the edge of their ontology is the work of "destrucktion," articulating the obsolescence of outmoded ontological categories. As Heidegger described it, "this hardened tradition must be loosened up, and the concealments which it has brought about dissolved" (1962 [1927], p. 44). Continuing on, these Creationist students exist with others in the process of education. I now turn to three primary categories of outlook toward evolution, and how these views interact to culturally produce contention around evolution.

Chapter 4
Evolution and Religion

The engine with murderous blood was damp
and was brilliantly lit with a brimstone lamp
An imp, for fuel, was shoveling bones
While the furnace rang with a thousand groans.

The boiler was filled with lager beer
And the devil himself was the engineer
The passengers were a most motley crew
Church member, atheist, Gentile, and Jew

Rich men in broadcloth, beggars in rags,
Handsome young ladies, and withered old hags
Yellow and black men, red, brown, and white
All chained together—O God, what a site!

While the train rushed on at an awful pace
The sulphurous fumes scorched their hands and face
Wider and wider the country grew
As faster and faster the engine flew

The Hell Bound Train (excerpts)
American Traditional

Creationists represent one of three general ontological positions regarding evolution. The 31 students I interviewed fit nicely into categories established by sociologist of religion Robert Wuthnow (2005), which I adapt slightly. Their stories spoke of similar religious experiences and perceptions of other religions respective of evolution. A small exception is the presence of slightly more agnostics and atheists than the seven or so percent described for the entirety of the nation (Edgell, Gerteis and Hartmann 2006). In general, my students came close to matching the religious identification percentages of the USA, as of 2008 (Pew Foundation 2008). Religious affiliation as a percentage of US total population was as follows (Pew data, then my data parenthetically). Christian total 78(77)%. Of the percentage of Christians, evangelical Christians 26(29)%, Roman Catholics 24(16)%, and other "mainline" Christians making up the rest. Outside Christian identity, there were the nonreligious 16(19)%, and 5(3)% percent of the nation practiced non-Christian religions.

D.E. Long, *Evolution and Religion in American Education: An Ethnography*,
Cultural Studies of Science Education 4, DOI 10.1007/978-94-007-1808-1_4,
© Springer Science+Business Media B.V. 2011

In interviews, students described their ontological position regarding evolution, science, and religion. Important to keep in mind for later, the categorical positions I discuss become analytical categories. I use these categories to explore the apparent stilted vocabulary that almost all students I interviewed held toward other religions vis-à-vis science and particularly evolution. With these three positions outlined, three students from each category were selected for an additional interview in which their life histories with science education, evolution, and religion would be more fully discussed. These case studies then are assembled here to provide the core to a polyphonic voice describing the ideology and commitments of these positions.

For Creationists, evolution education is not perceived as value free, certainly not free of affect, and in some cases, seems to be downright disturbing. Furthermore, as the sociocultural contingency of life dictates a good bit of the possible discourses that one might engage in the day-to-day, for many, evolution never really comes into dialectical circulation. Worldview is not something easily changed, especially by instrumentalist attempts. So then why would someone abandon the ideological commitments of Creationism in favor of an evolutionary view? I do not say this to strip away the possibility of human agency. I am simply recognizing the social fact that our identities and possible worlds within which we reside are highly structured by the idiosyncrasies and normative hand of culture. This should not be seen as bleak—for education can be one of the activities in life experience where organic growth toward possible rather than prescribed worlds can be fostered. This is the kind of development that Dewey at his best hoped for. But as I have stressed before, sociopolitical forces are at work which make fields of human practice not quite as open and as filled with possibility as one might think. Foucault certainly thought this way, and the last few decades' focus on critical pedagogy in educational theory has followed this path. But for our problem at hand, and those most invested in it, the natural sciences rarely seriously consider all the extramural forces which act upon people on their way toward science literacy. It is time to stop ignoring the cultural and embrace it—*for* the purpose of science education.

Insomuch as there continues to be a culturally reproduced tension between science and religion in the USA, I need to explore the nature of Americans' religious dispositions and how they affect one's receptivity toward evolution. Although it seems a bit dumb to say outright, Creationists are not in the game of rejecting evolution for the purpose of practicing better science. They are in the game of evangelism, *Truth*, and ultimate purpose. They would never claim this, but they are also agents of social control *par excellence*. On the face of it, this is somewhat of an exaggeration (especially if you do not happen to be a Creationist yourself), as few people in their everyday lives consider evolution. But for the sizable minority of Americans that hold ontological positions like Matt of the Campus Christians or Tyson, evolution is a symbol by which a battle of social worlds and ultimate ends is being fought. Evolution is simply off the table as a possible explanatory mechanism for those whose ontology deals with one perfect, unified, transcendental *Truth*.

Fig. 4.1 Scott's (2009) continuum of receptivity toward evolution

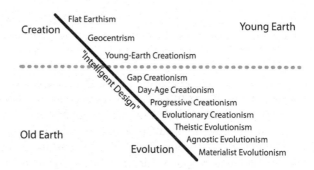

When evolution does get discussed in civic discourse, there emerge a handful of ontological positions that people disclose. The discursive interplay of these positions by people in social relation toward each other can describe how discussion of evolution is repelled, is avoided, or is fostered. These positions are greatly structured by the contingencies of the culture within which you are born, and work to structure the trajectory in culture in which you have moved. This is not to say that we are deterministically prescribed to *be* one way or another. This is to acknowledge that American culture continues to reproduce religious ontology of a few set types. We are not suddenly (excluding the obvious influences of immigration and global communication) producing vast numbers of Shinto or Hindus. Creationists *tend* to beget more Creationists. Theistic liberals *tend* to beget more of the same. How do these religious ideologies then affect the possible educational discourse regarding evolution? How do Creationists "get along" with others in educational discourse regarding evolution?

Prominent scholars of the evolution/Creationism "debates" such as Scott (2009) have provided a continuum by which we can see all the relative theological positions vis-à-vis evolution (Fig. 4.1). But this type of scale reduces an individual's religious identity to a descriptive label which, as I will demonstrate, does not do us much good when articulating the varying cultural commitments that ontological positions tend to prescribe for our discourses in practice. When and toward what ends do Young Earth Creationists, Theistic Evolutionists, and Material Evolutionists sit down in a spirit of open dialectic? We might better ask whether such a continuum says anything about the rationale of those who may be taught that this continuum exists, but simply shrug and continue on in their business.

The trouble with such a model begins with the categorical fixity that one might be tempted to read into it, and what it might imply about religious disposition. Problems with the utility of this model became clear to me when, during the early stage of this research, a student explained to me that "she's a Catholic, but goes with her friends here [at college] to a Southern Baptist church." The rub is that the doctrinal position of the Catholic Church she attended in youth is in recent decades plainly accepting of evolution, although she was a bit skeptical. Additionally, the Southern Baptist Church, part of the larger institutional force of the Southern Baptist Convention, is openly hostile toward the concept. So which one of her religious practices toward evolution is "right?" Which position has "misconceptions?" Past these muddying issues of identity, epistemology, and practice, what then is the use of such a tool above when actual practices tend to be much messier or even contradictory?

There are ways around this kind of problem. Sociologist of religion Robert Wuthnow (2005, 2009) provides a useful framework with which we can consider the effects of how people *use* their religious commitments, and how this effects evolution education. Speaking about the experiences of Americans from a large-scale study of how people work with or against increasing religious diversity on the USA, Wuthnow proposes thinking about religion and science debates in terms of the clarity of more general, but perhaps more bluntly accurate, ontological positions disclosed in actual social practices. By doing this, additional denominational messiness can be sidestepped when thinking about the negotiations that people make with science and religion. For example, consider Wuthnow's questioning of the institutional purity of either science or religion:

> If science and religion are distinct institutions, then this is…reason to believe that potential conflict can be overcome by recognizing differences. These are not merely conceptual differences, such as one emphasizing the natural or the supernatural, but also of social arrangements themselves (p. 160).

For those lacking the inclination to look upon the world sociologically, religious disposition then is not a matter of a randomized sprinkling of people who have "chosen" their religious faith. In this view, the hand of socialization trumps all other forces. As Wuthnow further clarifies by analogy, this view is misleading in that it misrepresents the nature of the epistemological commitments within the ontological business at hand:

> A baker and a butcher might be mutually dependent, but it is less clear that science and religion bear the same relationship to one another. A better analogy would be two all-purpose delis, one specializing in bread and the other in meat, but both seeking to expand. Science and religion come into conflict because neither stays neatly in its respective sphere (p. 160).

By this assessment, views about science like Gould's (1999) "non-overlapping magisterium" appear to be in trouble. It does work for some people, but not all people are willing to make the ontological commitment the view entails. Similarly to the Platonic farce of stripping affect out of everyday reason, both scientific and religious knowledge affect our ontological positions and the decisions we make from them. Wuthnow works from the position that social practices and attitudes toward others using religious commitments are more telling for untangling the American relationship toward both. Rather than inscribe people as passive actors carrying out the commitments of either science or religion or variably both, Wuthnow (2005) works with three categories onto which I will map my ethnographic data.

For Wuthnow, he sees American religious practitioners made up of spiritual "exclusivists," spiritual "shoppers," and "inclusivists" (which I will treat as a singular, but compound class), and an additional minority category of agnostics and atheists. Spiritual exclusivists are hallmarked by their perceived access to the philosophical absolute, that their worldview cannot possibly be wrong, and all others are deficient—a product of a "fallen" world. As we have just seen, contemplating evolution for this group can prompt existential anxiety. Here, we examine the position generally.

Spiritual inclusivists might be seen as a broad category of any theologically liberally minded, ecumenically inclined religious practitioner—ready to go along (to quite great lengths) to get along. Spiritual shoppers are a particularly higher educated, well-traveled, and less deeply theologically committed strain of these inclusivists. Lastly, although Wuthnow does not focus much on them, the agnostics and atheists within American society make up a stable if small minority. Although there are ontological similarities between the agnostic and atheists and the spiritual exclusivists—when we consider the nuance and distinction by which these arguably polarized forms of the same ontological type look upon science, we must hold them apart for clear sociostructural reasons which I will detail.

4.1 Position I: Thou Shalt Have No Other Gods or Epistemologies Before Me

Let us work with this central edict. For students who hold this ontological position, evolution, quite simply, cannot ever be true. As Hannah plainly put it: "We're Christians so it's kind of like...we didn't evolve from monkeys." This is the epistemology of the prototypical Creationist. For evolution to be true requires undermining absolutist conceptions of *Truth*, and the source by which such *Truth* is derived. As was common in Creationist student conceptions of evolution, human evolution or frankly any macroevolutionary process was impossible to conceive of. As Randy explained regarding human evolution: "the evolution of apes to humans... it doesn't really like...*focus in*...because if it did, then we'd still have more apes becoming humans today." As I have discussed elsewhere, ontologically, one cannot be a Creationist and be prepared to accept evolution (Long 2010b). Of the 31 students I interviewed, nine identified themselves as a Creationist, six of the "Young Earth" variety, and three who accepted evolution for nonhumans but insisted on special creation for humans. All described a religious position respective of Wuthnow's religious exclusivist typology within which they would nicely fit.

Creationist students of Protestant denominations were all strongly evangelical—Pentecostals, Church of Christ, and Southern Baptist. "Strongly" being those who actively look to convert others. Similar to the theological outlook of Wuthnow's Christian exclusivists, the Creationist students I interviewed saw their religious constitution as *the* one true way. Although going in to my interviews I did not expect it, there emerged a striking Foucauldian rhetorical power play at work in the discourse of Creationists. In many cases, when I asked students to identify their religious identity among others, a synecdochic claim to "just Christian" was firmly stated. This move was specific to Creationist students. As Andrea typified the position, "I don't really associate myself with a specific denomination. I just am a Christian, end of story." This type of distinction appears unique to American evangelical Christianity and the cultural bleed-over effects this movement has had on other Christian denominations.

Although science educators writing about Creationism have discussed the intended benefits that historicizing epistemologies while confronting Creationist resistance (e.g., Settelmaier 2010 and Alexakos 2010), it is actually the epistemological exclusivity of the exclusivist position that forecloses thinking more broadly. Typical of the neo-Gnosticism that Bloom (1992) sees distinctive within American religion—the Southern Baptists and the Latter Day Saints—these students had precious little conception of their faith in historical or sociological context. As Tyson said best, "I don't understand denominations completely myself...We focus on *individuals* taking the Bible. The Bible is where we take our cues from. That's what we go for. We want it to affect our everyday lives."

But what then is the basis in practice for the rhetorical claim of having *the* way to God? Why the insistence on such a limited repertoire of theological influence? As it seems, for many, there is little opportunity to examine other paths. As for difficult ontological questions and for the comparatively more mundane exposition of evolution—both appear to be conceptually outside of the epistemological repertoire of Creationists. In social practice, these types of discussions might never have opportunity to come up for fathomable reasons. Although one could boil this exclusivity versus inclusivity down to basic social intolerance versus tolerance, Wuthnow points out that both actually share a similarity of sorts. "Christians who believe that only Christians are saved also share some of the views that permit inclusive Christians to sidestep some of the hard questions this belief entails" (Wuthnow 2005, p. 180).

As becomes clear when exclusivist Christians "stand out" in a way the inclusivists do not, regarding their views on evolution, exclusivists "recognize that what they believe is difficult to defend without seeming arrogant, overconfident, and irrational" (Wuthnow 2005, p. 186). This is compelling as irrationality is usually the number one pejorative that vocal natural scientists foist upon Creationists. In this way, some Creationists, while attempting to acquire the social capital of university degrees in "secular" institutions, have learned to keep their mouths shut. Renee discussed reaching out to another peer while studying for a biology exam:

> One girl was wearing a necklace which I thought was a cross, but I guess it was kind of like more of an X, like made out of ribbon. So I thought it was a cross and misinterpreted I guess. So as we were studying for a biology exam, I brought up Creationism. And she said, 'Well, yeah, I guess technically evolution is a belief like Creationism'. And she kind of laughed about it and that was it. I just thought oh, okay, I'm obviously off base.

For some others such as Hannah, her reaction to this kind of challenge was a bit blunter. Our interview vignette shows this:

> This girl told me I evolved from a fish and I told her I didn't.
>
> *Well, why did that conversation come up?*
>
> Well, I don't know. She was just talking about it one day in government class and was saying how I evolved from a tadpole. And I'm like, '*You're out of your mind*'. No offense to her, but I did not evolve from a tadpole.

Traditional social power and rigidly prescribed gender scripts are also usually part and parcel of this type of evangelical practice. As was clear when I examined the organizational structure of each of these churches, each one was entirely made up of older white men, although women did show up on social committees and in the day care departments. Again nodding to the issue of social power and control these types of institutions levy on this category of people, as Wuthnow explains: "This learning style emphasizes trusting in an authority figure or authoritative text for answers, submitting graciously to what the Bible, the pastor, one's boss, or one's spouse may say is right" (p. 169). Typical of students in this social category, evolution posed problems for their faith's authority in that it forced a decision between competing ontical claims. Either Biblical literalism was wrong or somehow the science supportive of evolution could not quite be right. As already evidenced in the existential anxiety a good number of these students felt, the "learning" that mattered when these difficult topics and competing interpretations arose was that of the home and church. As Wuthnow asserts: "Exclusive Christians are more likely to believe in a single set of right answers, which a person questions or tampers with at his or her peril" (p. 170).

Although the Christian exclusivist position appears dominated by strongly evangelical Protestants, as my interviews clarified, they are by no means the only people demonstrative of such views. Mary, who was raised as a United Methodist but later converted to Catholicism, had similar views: "I grew up Methodist, so that's where most of my background has come from. I feel like evolution conflicts with the religious beliefs I have. I believe God took seven days to do it, and that's in Genesis." Unpacking what appeared to be deference to the type of patriarchy described above, Mary explained her religious conversion and the social influences on her regarding scientific matters:

> I changed faith practices when I was about 20. I changed because I was getting married, and my husband is Catholic… I'd say I probably got a lot of what I believe about evolution from my dad. He's probably said it whether or not I realize it. I don't specifically remember him saying anything, but my mom has never been really outspoken. A lot of what I have known from growing up, I'll probably find somewhere in my life that my father is the one who said it.

In an interesting demonstration of exclusivist Christian theology in practice, when Mary describes the larger family discourse regarding evolution, general intra-exclusivist tensions arise. "My brother is really religious, but he and I don't talk about it that much. He goes to a Church of Christ that is a very anti-Roman Catholic church, so we kind of bash [sic] a little bit on that."

As a Christian science educationalist writing from outside the USA, Settelmaier's (2010) work illustrates how the distinctiveness of these culturally specific nuances within the USA could be lost on an outsider. In fact, it is an indicative hallmark of the recent evangelical movement, when pushed, to rhetorically marginalize "denominational" Christians as non-Christians—that is, for Protestant exclusivists, Roman Catholics are not Christians. I learned this myself as a child in heavily Scots-Irish and German Protestant-influenced southeastern Pennsylvania. On more than one occasion, slights were directed by friends' parents toward the small minority of "Papist idolaters" within our region.

So what stops Creationists from learning about or considering more theologically moderate positions that do not find scientific knowledge contrary to the teachings of the Bible? What is the role of education (both scientific and religious) in this scenario? Seemingly, every step down an educational road supportive of evolution involves moves that dissolve the ontical distinctiveness of absolutist ideology. This is certainly not to say that people do not make these moves. Some certainly do. But one does not hold an absolutist ontology and begin to consider matters of interpretation, historical contingency, and social power. *Truth* is, for such people, timeless and immutable. To question this type of *Truth* meaningfully is to shake off the spell that such *Truth* holds over its believers.

As Wuthnow (2005) describes, the social arrangements of exclusivists' lives and faith practice actually explains the strength and reinforcing structure of such congregations:

> Exclusive Christianity is nurtured in exclusive churches where people…learn early that they are in danger of eternal damnation and accept Jesus as their savior to avoid suffering eternal damnation. They learn that many people outside of their own group—certainly atheists and followers of other religions but also many self-proclaimed Christians who have not truly turned their hearts to God—will not go to heaven (pp. 164–165).

Tyson described his own experience with the slightly unsettling prospect of coming to a public college after having been homeschooled. As he explained, his college-age church youth group draws encouragement from each other as they enter the "world." He calls this process "breathing," in that he goes out into the world he is in but not of, and retracts to the congregation for assurance and support. As "salt and light" as he claims, he is compelled to go back out into it. As was clear in Tyson's earlier tensions in discussion with his mother about feeling "torn," Tyson is affected by the epistemological temptations of the "world" but has not of yet dabbled in aligning with it. Again describing the mediating influence of youth social groups, another Creationist student named Hannah discussed her experience going to the Creation Museum. "We just talked about it one day. Because we went to the Creation Museum and we just discussed evolution…why *some* people think we evolved from monkeys and what the church believes. And so it got cleared up."

Andrea (who attends the same Southern Baptist church as a prominent Creationist leader, and is getting dually certified as a biology and chemistry teacher) additionally adds her perspective on this same influence as she weighs scientific "evidence":

> I am really involved in my church, and I have been really blessed with some very intelligent staff there, and whenever we talk about origins of the earth and whatnot in church, there is always a blend of science …these scientists have said this might be a possibility according to what the Bible states. I've also gone to the Creation Museum several times, which I think is really awesome because you can go through there and it shows you the different places where evolution doesn't really have a specific answer for things…My favorite part is where it talks about the Grand Canyon, and it shows you how the Grand Canyon could have been made in such a way that it would have only taken a couple hundred years. One of the examples they give is Niagara Falls. [It] loses inches and inches of its…I don't know— where the water plummets, I guess the cliff, each year. It loses inches and inches off of that every year, and at that kind of rate, it would have only taken several hundred years for the Grand Canyon to be formed. So why couldn't that have happened?

As a general statement lacking in the detail of Andrea's comments, but typical in the reasoning of Creationists I interviewed, Randy talks about how he deals with science supportive of evolution in class and how this affects his discussions at home. "My mom and dad don't really believe in it. They take the same view on it as we do. It's material that we're supposed to know for a class, but it's nothing we're supposed to take to heart."

Rejection of evolution was the prime organizing principle by which this category of students was selected, but is that all they share in common? What other qualities were indicative of this group? One fortuitous turn was when Tyson disclosed the rich history of his homeschooling experience. As an instructive extension into the worlds of these students, the homeschool movement is dominated (at least in the region of Mason-Dixon State) by strongly evangelical Christians, and professes a fairly clear ideological line against the kind of intellectual ground in which evolution is usually supported. As this specific groups' charter makes clear, with homeschooling "parents can control destructive influences such as various temptations, false teachings (including secular humanism and occult influences of the New Age movement), negative peer pressure, and unsafe environments"...and as a goal will "protect children from mental, physical, emotional, and sexual abuse by secular humanists in a socialist society or governmental system." Respective distance set aside, what does this type of ideological line usually cover up? As Wuthnow (2005) has pinned down reflective of the subtext of this movement:

> It is not uncommon for exclusive Christians to associate the growth in diversity with some frightening, apocalyptic vision of the end times...The threat is more likely to be cast in terms of homosexuality, promiscuous lifestyles, or relativistic values being taught in public schools, any of which may be loosely associated in people's minds with diversity (p. 184).

As many who have grown up in this tradition can attest, the "Devil's work" is embodied by anyone attesting to being a Secular Humanist—for this is to take on the epistemological mantle of the "fallen world." Regarding the sociological room for potential racial exclusivity, a spot check of Creationist student church congregations found them to be exclusively white assemblies.

As the historical racism associated with such groups has become a political liability, Answers in Genesis (who runs the Creation Museum) has turned toward portraying evolution as the means by which all American racism has been fostered. In lengthy displays, evolution is associated with social Darwinism. Exclusivist Christian congregations are almost unanimously anti-gay. Although it is not explicit in museum displays, pushing back on social "evils" such as homosexuality are a prime fear and rallying point for exclusivist Christians. The stark difference of this epistemology versus a more moderate one is evidenced by the phenomena for which an "evolutionary worldview" is blamed (Fig. 4.2).

The worldview these students' congregations and social networks work to uphold is one, whether rightly or wrongly, seen by them as being under attack. I have so far refrained from explicit political analysis, as I will discuss this thematically in a later chapter, but it is worth mentioning in brief. When asked, all but one Creationist student identified as a member of the Republican Party. More interestingly, when asked whether they believed that one's political affiliation had any relationship with

Fig. 4.2 Typical of the exclusivist Christian worldview, famine, animal predation, nuclear war, genocide, pain in childbirth, violent weather, and drug use are suggested as results of an "evolutionary worldview" at the Creation Museum

their opinion regarding evolution, all but two saw no relationship. When these two students hedged, Andrea's response typified the view:

Do you feel that a person's political affiliation has anything to do with their views about the topics we just discussed?

> I'm going to say no only because I feel—I guess it could, but I really feel like people should choose their political party based upon their own beliefs and not believe just because they're in a political party.

Apparently, recognition of how we share social commonalities with norms of one's political movement is lost on a good many students (and certainly many, many Americans in general). Andrea's reality is influenced heavily by *should*—"people should choose." Contrariwise for this project's other more theologically moderate or secular students, whether identifying as Republican, Democrat, or other these students almost unanimously identified the relationship between those that resist evolution and politics. Responses to this question were often identical when asked—"the *extreme* religious people" or "the religious conservatives." As Apple (2006) nicely ties this altogether in its implication for evolution educations future:

> The authoritarian populist religious right believes they are under attack. Their traditions are disrespected; the very basis of their understanding of the world is threatened. Evolutionary perspectives are not simply one element among many in the curriculum that wrong them. Such perspectives go to the core of their universe, even though they may not fully understand their own history in regards to the positions that they take (p. 134).

4.2 Position II: The Both/and

Our second and largest category is that which Wuthnow (2009) describes as the "Both/and" position. This is the type of person inclined to variably see the merits of *both* religion *and* science, rather than the more absolute and final ontology typical of Creationists, and scientists for whom natural science represents the means by which all questions of any category are appropriately answered. Students within this Both/and position toward evolution were a majority, similar in percentage to their adult counterparts at large within the USA. The criteria by which Wuthnow and others, such as Marty (1998), ask us to rethink about what our religious commitments mean give us a picture that is not as clear or unproblematic as it might first seem. As Wuthnow (2005) explains, what is important to understand about the American relationship between religion and science is the fact that the problem is not actually worse. Rather than focus on the fringe elements of this relationship, how is it that so many people in fact see no problem? This is important once one remembers that the main players in constructing such tension are most often active or powerful interests within either the scientific or Creationist communities. It is also the case that these interest groups are essentially composed of either religious exclusivists (in the case of Creationists) or religiously unaffiliated or outwardly atheist scientists.

A central issue for Wuthnow (2005), which is of no less importance to this project, is how the following seeming slight contradiction is adjudicated in social practice regarding evolution and Creationism: "In one national survey 57% of churchgoing Christians said it was not only true that 'Christianity is the best way to understand God' but also that 'All religions are equally good ways of knowing about God'" (p. 131). The obvious extension we can draw has direct implications for how we can think about the public's attitudes toward evolution/Creationism debates. How is it that, against the better wishes of scientists pointing out the lacking evidence supporting Creationist "science," that most Americans feel disinclined to push back more strongly against it? As Wuthnow sees it, "discussions [like these] are difficult…not because the public is caught up in bitter disputes about what should be taught in science classes but because the Both/and view of science and religion has become all too familiar" (2009, p. 177). Like the normativity of Christianity *over* all other religions and the nonreligious in the USA, this ecumenical balance between science and religion, rather than only religion or only science, has a loose hegemonic control over most polite social discourse. For those so inclined, radical or fringe views then are to be avoided, as they represent being too far out of line with what "one" does.

Understanding where natural scientists usually stand on this issue helps to clarify things, and is worth quoting at length:

> [A]…factor that aggravates the potential conflict between religion and science is that relatively few scientists are religious (at least in conventional ways). For instance, in a recent national survey of physicists, chemists, and biologists at elite research universities, only 8 percent said they had no doubts about God's existence, while 38 percent said they did not believe in God, and another 29 percent said they did not know if there is a God and believed there was no way to find out. In the same survey, 55 percent said they had no religious affiliation and only 16 percent attended religious services at least once a month. These figures underscore the sharp difference that exists between scientists and the general public, where, for instance, only 7 percent can be regarded as atheists or agnostics. (Wuthnow 2009, pp. 166–167)

Ecklund's (2010) study of research university scientists' attitudes about faith finds much the same, but adds the note that scientists often lack an appropriate vocabulary to discuss matters of faith:

> While elite scientists have a very elaborate vocabulary for the subjects they deal with in their own particular field and subfields, those without religious identity (over 50 percent) have limited experience or limited ongoing interaction with religion and religious people (p. 133).

Contrasting this with the vast majority of the public for whom doubts about the existence of a God are few and far between, the potential for a disconnect between the two is unsurprising.

So, given that the hopes and fears of scientistic scientists and Creationists are so often publicly forthright, how would most students contrast this? As was similar to the types of epistemologically open but non-committal people Wuthnow (2009) discusses, these students also were uniform in their commitment to not having *the* answers. Allison considered why there is social conflict between evolution and some religious faith. "I think that in science they try to prove all the things that happen, but some people believe it differently…I think it's probably a good idea but people still can believe what they want to believe." Mitch, a premed biology major echoes this non-committal stance, and makes the common move that some noted theistic evolutionist scientists such as Miller (2008) have worked hard to quash. "It [evolution] can really only be described as a theory…some will argue that there's concrete proof, some will argue that there isn't." One might give Mitch the credit that he understands the conceptual rigor that formal scientific theory entails, but he deflates this hope when I asked him to consider the possibility of teaching Creationism or Intelligent Design alongside orthodox science in the public school classroom. "I would not be opposed to it…I think both theories are important to the educational process…I think providing options is the best way to do it."

Chad's response, while a bit less articulate, still carried the same message: "It's something you can look into…I don't know who's right or who's wrong. I just want my kids to…as a father you need to get all the information, you get everybody's sides, and then you decide what to do." Chad's case was especially telling as he made these claims in an interview directly following a few days of explicit discussions on the history and nature of science behind evolutionary theory—but, as he also noted: "I've never really been into science…it was my worst subject in high school." Conversely, this isn't to say that Chad is strongly religious—quite the contrary:

> I was born Catholic and all that…and I was told that God did all this stuff, but as I got older I didn't really buy into that anymore. It kind of makes sense, I kind of lean toward evolution and maybe it is true that everything just happened. I'm just open…but I'm not an atheist because I do believe in a God.

Of this tolerance for "all the theories," a side effect of postmodernism's central ethos, Christian agrees. "I wouldn't have a problem with it…I think it's good to look at different points of view and ways of looking at things," as does Joe. "I would be fine with it. I think they should know about it." This issue, the wishy-washy tolerance for Creationism and Intelligent Design's creep into the public school curriculum, is a flash point for concern for most biologists. It was also a thematic hallmark and point of empirical ubiquity for each of these Both/and students, which I will discuss at length in Chap. 7.

For some Both/and students even bringing up evolution, and the social discord it has prompted in the USA, elicits a bit of edginess. As Shelly immediately clarifies when asked, "I *do* believe in evolution... I do see the evidence for it...but I *don't* believe that just because you believe in evolution doesn't mean that you can't believe in God." The terseness of Shelly's reply belied her experience with evolution having prompted tension in prior social situations. While discussing a conversation that she and a friend had after evolution was taught in her high school class, the rhetoric turned polemic. "Evolution is a load of crap" Shelly reports her friend stating. Continuing in summary of this friend "she believes that the Bible is the world of God and that's it." Shelly interprets this view of religion as a problem (*with a tone of slight disgust*):

> People think that just because...their first thought is like *how could we come from monkeys...* they don't really...they think that by saying we're related to monkeys that it's like 'poof'—monkeys to man...and it's they don't get that it's the case that we *slowly* evolved... they take the Bible literally, just word-for-word.

This view was common to these students. As Wuthnow (2005) typifies of this relation toward religion:

> They often make a point of saying that they interpret the Bible metaphorically rather than literally—an argument that distances them from fundamentalists but the significance of which is missed if this is all it is interpreted to mean. The key is that inclusive Christians interpose an interpretive step between the context in which the Bible was written and the words that appear in the Bible (pp. 146–147).

This issue for Wuthnow, Shelly, and her ontological peers cannot be stressed enough. It is not a new insight regarding conflicts over evolution in the classroom, but the full implications of what it would take to overcome this have not yet been adequately addressed by anyone I have encountered, save Toumey (1994). Why does "one" take the Bible "literally," or why does "one" approach it metaphorically? Why would "one" be inclined to "believe" at all? Lest we have the deterministic belief that Christian fundamentalist congregations are clusters of cognitively "under-developed" people, the answer we are looking for is one of socialization into an epistemological culture.

As was common to about half the Both/and students I interviewed, the fact that evolution itself was taught was not an issue for them *per se*, but rather the social tension itself became an issue. As Mitch relayed, "in high school people would see that evolution caused some people to react and it would start debate...and other students would want to know why someone would have such a strong reaction to it." Mitch had a similar experience and feelings:

> In eighth grade there was a growing debate in our school about the legitimacy of evolution...some of the more devout Catholic parents were trying to press the issue of Creationism, the other side of course was for more evolution based learning...it grew to divide the class a little bit but I'm not sure how many of the students actually cared that much about it.

This kind of recognition of social tension, colored by what appeared to be a wish for a less contentious tone was indicative of the Both/and students. Not all saw this, in fact a few seemed quite oblivious to the whole affair (and for that matter *many* other things). But the clarity of this view was clearly at odds with the Christian exclusivist students—who very much felt as though they were being marginalized or slighted. "I wish they'd teach both sides"…"they could show the other side." This dichotomous thinking was never violated by Creationist students.

So what of the social interplay of these two rhetorics—the inclusive and the exclusive? Why, if students and teachers of the Both/and position dominate the civic landscape, is there not a more ecumenical dialectic supportive of evolution in everyday educational practice? The exceptions are there—very good ones which I will discuss in Chap. 7. But what drives the otherwise leveled-down commitment to discussing evolution? As some Both/and students show in their relation toward their own faith, unlike Creationists for whom evolution was often a topic of (critical) discussion at church, Allison states: "We haven't really talked about it at church." This unto itself is not surprising, but buried within Allison's otherwise inclusively tolerant views sits some doubts. She hedges regarding evolution: "There's some parts that I just kind of question…that I'm not sure…when we were talking about it my teacher…just the whole thing sometimes…I thought that maybe some of it had to be true but maybe not other parts…." When I ask her to clarify what drives the uncertainty of her disconnected phrases: "I think it's because of my beliefs…I think that some of this might not be right."

This type of view is not new to me, as the earlier Catholic who attends Southern Baptist services with her friends reconfirms. But typical of most Both/and students, many of them have little or no idea that their own religious community might be okay with evolution. Often, it seems that some of these folks have a more theologically conservative view of their own church's doctrine due (possibly) to the general cultural influence of loud fundamentalists. Mitch, who chooses to live the ecumenical line, both recognizes what he perceives to be a conflict and works his own way past this. When I asked him if the teachings of his religion conflict with evolution: "It certainly does…my religion believes that a higher power created life as it exists today…of course evolution explains life…how it evolved from one stage to another….and the idea of that in the eyes of the religious person that I used to be would be that it takes away the divinity of the creation—so it's not a popular topic with Catholics." As he drew out his own distance from what I take to be his discomfort with this, he described having support at home. "My parents are pretty [much] at ease with balancing the idea of evolution with the idea of creation." All throughout Mitch's interview, characterized by his calming, introspective tone I got the impression of someone very comfortable in his skin. Wuthnow (2005) summarizes this position:

> Maintaining a commitment to Christianity while holding an open-minded view of other religions is, in the final analysis, a matter of arriving at a delicate balance between what one may expect to learn from other religions and one's reasons for preferring a single religion (p. 152).

Although Mitch saw problems with what he took to be the orthodox Catholic line regarding evolution, the overarching commitment, support, and stability in his life lent little motive to distance himself from incongruities of faith, unlike some in my final category of students.

4.3   Position III: Agnostics, Atheists and the Nontheological

Although exclusivist Christian Creationist groups such as Answers in Genesis have attempted to typify mainstream science as supported by or leading to atheism, the actual picture of social identification as a nonbeliever appears neither that profound, nor, as some like Answers see it, sinister. There are simply not many nonbelievers in the USA. But of that small number, it also happens that for our contention between religion and science over evolution, a goodly number of those most inclined to act politically against antievolutionary forces are nonbelieving scientists. Additionally compounding this, as discussed above, natural scientists (at least at research institutions) are disproportionately nonbelievers.

The remaining six students I interviewed fit within a general category, which I will refer to as atheists and agnostics. Of these six, two identified as atheists, two as agnostics, one as a nontheist, and one as a transcendentalist. As members of a social category that appears poorly understood and has little institutional representation aside from academe (where the view, at least in the natural sciences, is common), what can we say about this group? Perhaps due to the relatively small amount of these people upon the American stage (at most 7%), Wuthnow (2005) gives them no special category. For my purposes, and given the nature of the evolution/Creationism debate sociologically, I must. As we have already seen, although some natural scientists may muddy the issue by having "unconventional" views, the issue for Americans at large toward this group is one of skepticism and in some cases distrust. Wuthnow (2009) contextualizes this: "The high esteem in which the public holds science and scientists, if believed, is tempered with the populist ambivalence that colors public attitudes toward all elites. Admiration bleeds easily into envy and further into disdain" (p. 166). This is no small point when we consider that this first assertion reflects the public's attitude toward science as a general enterprise, not specifically the populist reception to what natural science has had to say regarding the origins of humans. Add the complicating issue of origins *into* the prior discussion of exclusive and inclusive Christian discourse, and the state of tensions is yet clearer.

Although being a nonbeliever in the setting of the natural sciences within the confines of the academy may put one at little risk of social retribution, when moved to general cultural discourse, the relation changes. Atheists and agnostics, at least by the terms of exclusivists Christians, are "bound for hell." Reflecting the seeming difficulty to get a feel for who comprises this social group (except for members of this group), Edgell et al. (2006) instead assert a Durkheimian notion: one best understands a group (nonbelievers) by the social attitudes believers have toward them. "Atheists are at the top of the list of groups that Americans find problematic in both public and private life, and the gap between acceptance of atheists and acceptance of other racial and religious minorities is large and persistent" (p. 230). As they discuss, this gap is greater than that tolerant of queer identity, or frankly any other symbolic identity their survey measured. This likely implies that there are actually more agnostics and atheists in the USA, who claim a conventional religious identity to avoid conflict. Importantly for our project here, these social conceptions were not usually borne of "knowing" an atheist, but rather "acceptance or rejection of atheists is related not only to personal religiosity but also to one's exposure to diversity and to one's social and political value orientations" (p. 230). Both of which are of prime importance as we have seen, and will continue to see, regarding relationships toward evolution. Given the social conditions that shape religious exclusivists possible daily discourses by the people that they do (or do not) meet, religious exclusivists are highly unlikely to know atheists well.

Of the students I met who fit this bill, what can I say about them? The strongest thematic claims I can make about them, based on the distinctiveness of these groups' narratives, was their authoritative speech toward science, and how their relation toward evolution involved social conflict. They, unlike the other groups, had life experiences in discussing evolution which for them involved anger or pejoratives. Whereas Creationists describe alienation at the hands of "liberal" or "atheist scientists" who "don't present both sides," agnostic and atheist students were resentful or angry at the social impact of Creationism on science. Additionally, their discussion of evolution prompted anger in some Christian exclusivists—often family members. The second thematic aspect distinctive of this group was the clarity in which they spoke of what evolution, and for that matter the nature of science in general, is. Whereas I will rule Creationists by the merit of their inclusion of Creation "science" as reasonable science out of a clear understanding of the nature of science; atheist and agnostic students often made clear distinctions respective of the current methodological orthodoxy of rejecting teleology when *doing* science. The Both/and students, for whom science was often a calling or an interest, spoke with far less uniform clarity about science.

James, who had perhaps the most direct expression of frustration with Creationists, pointedly stated the problems he sees with discussing Creationism or Intelligent Design in a science class. "I'd be rather angry because that's not science, that's faith, there's a difference." Sheena, following a clear church/state separation line, expressed similar feelings:

> I would not like it. I don't feel it's the school's right to push religion and things like that…
> if I want to teach my children religion and stuff like that I'll take them to church, they don't
> need to learn that in school. There's more important things that should be focused on.

Will, a soft-spoken premed student quietly stated the same sentiment. "Just give the facts…leave religious or other opinions outside." Speaking about his perception of how Creationists, and for that matter the general public, think about science and evolution , Donald lays out a frank assessment of the nature of the conflict:

> There's this idea out there that when you say 'theory' that it's just something that someone
> pulls out of the air…or random notions that someone just puts things randomly together -
> like it's all just a bunch of bullshit. There's no sense of when you say 'theory' that there's a
> whole process of the scientific method.

Sheena, who indicated having advocated for evolution many times in her life, described her rhetorical tactic:

> When I hear evolution being brought up I always think of the common misconceptions of
> turning from monkeys to people. I think the alternative to evolution is absolutely ridiculous.
> I'm not a religious person, so, when people ask me about evolution, I always seem to think
> about the opposite…like I don't usually…how should a say this…I don't back up my side,
> I usually argue against the other side.

From the directness of Sheena's advocacy, there were also cases where discussing evolution from these students' positions could prompt conflict. In ways, this was also common to the Creationist students I spoke with, but in this case in an inverse relation. Tracy shared how he "learns" what one discusses and what one does not. "I just remember a specific time when if that topic was brought up—that was a very, very bad topic... I just learned that when we brought it up it was a topic you just didn't discuss." This exchange took place between himself and his girlfriend's parents. When I asked him whether this was his first time experiencing this tension, he confirmed: "That's when I pretty much found out. They were very angry about it... it was very heated and anxious."

The atheist and agnostic students were distinctive by the clarity, interest, and authoritative voice they used when speaking about science. James shared his interests with enthusiasm. "I like physics. I like biology. Chemistry is interesting; anatomy is very, very interesting to me. I don't have a lot of education outside of high school in it, but I'm constantly watching the science Channel, and Discovery Channel, and all those things." Sheena echoed James' enthusiasm in sharing what she likes about science:

> Everything. I've always been a science buff. I've loved all my biology classes. I took an anatomy class in high school and it was probably the coolest class I've ever taken. I love chemistry because I love the way they can just break everything down to just fundamental aspects and how they break down just everything on earth.

Tracy probed into conceptual territory no student outside this category attempted. "I really enjoy quantum mechanics. I also really enjoy biology a lot because of aspects of the way it looks at life—looking at life with a more focused eye...looking at the smaller parts that make up a larger part." Tracy continued by making a type of distinction exclusive to these students, and explaining his interest in quantum mechanics. "I'm interested in philosophy, so I think that plays into that...looking at life scientifically and then looking at life in a different sense." Whereas the Both/and students might not object to human evolution and Creationists certainly did, Will typifies the comfort and terminological adeptness that these students had in discussing human evolution. Explaining to me what evolution is: "It's the theory that man evolved to what he is, from Cro-Magnon up to *Homo Sapiens*."

The three ontological positions toward evolution I have detailed will now follow us along as I continue to take us deeper into the social world of evolution education in practice. In the next chapter, I will discuss the conditions by which evolution becomes reasonable for a person where once it did not. Past issues of existential anxiety, how does someone who abandons a Creationist world fare in their everyday dealings? Flying in the face of "misconceptions" and "conceptual change theories" of science education, acceptance of evolution is to be socialized to a scientific world in which evolution is *needed* as a conceptual tool. The existential anxiety and categorical framework that I have described are a somewhat static view of dispositions toward evolution in culture. Using Anne Swidler's (1986) "strategies of action," I will now animate students' worldviews in action—those who are changing ontologies in the midst of their educational experience at Mason-Dixon State.

Chapter 5
Evolution and the Structure of Worldview Change

Run to the moon: O moon, won't you hide me?
The Lord said: O sinner-man, the moon'll be a-bleeding

Run to the stars: O stars, won't you hide me?
The Lord said: O sinner-man, the stars'll be a-falling

Run to the sea: O sea, won't you hide me?
The Lord said: O sinner-man, the sea'll be a-sinking

Run to the rocks: O rocks, won't you hide me?
The Lord said: O sinner-man, the rocks'll be a-rolling

All on that day

Sinner Man (excerpts)
American Traditional

5.1 How Evolution Fares When Worldview Changes

It is worth reflecting upon where I started and where we have been. In the prior chapter, I identified three main ontological positions which when interrelated affect the types of discourse toward evolution that people tend to have in education. It became increasingly interesting to me what exactly might prompt a major shift in student views toward evolution, given the anxieties, perceived fixity of their worlds, and rutted traditions at stake in some of these students' lives. So how in fact might someone dramatically change their position toward evolution? As it is likely that acceptance of evolution is but one marker of a larger ideological whole, what does it take for someone to make this switch? Given the existential vertigo that Creationist students were experiencing, what might prompt a student to fall (or get pushed) off this precipice, and where might they land? The "scientific habits of mind" discourse within the AAAS *Project 2061* goals speak clearly about the preferable adoption of a scientifically literate worldview by students, but nowhere within this or other similar documents is there a serious consideration of what other ancillary phenomena might be taking place along the way.

In Chap. 2, I discussed the concept of worldview, the trajectory of its invocation through philosophy and into twentieth century social theory, and how it has shown up in the narrative of an unabashed, but clearly worried, Creationist student. At the same time, science educationalists have begun to consider worldview. This second point is of recent concern, as, while educationalists consider whether or not to operationalize a term, Creationists have already moved to articulate worldview as a tool against evolution and some whole fields of science. Tyson, who earlier deployed worldview when I asked him about evolution, sat down with me to speak specifically about how he came to frame his understanding this way. This time more resolute, Tyson expressed none of the hesitancy of our first discussion where he expressed "feeling broken":

> I learned about worldview from a gentleman named Dr. Noebel, he wrote 'The Battle for the Mind'. I derive a lot of my thinking, from him. He had a thing called "summit" out in Colorado, and I went to that. He has a whole chapter where he lays out all the worldviews... the cosmic humanist, secular humanist, Christian, Marxist, and then there's the...also the Islamic worldview which he added to his newest book. In that one, there's not really an evolutionary worldview...but cosmic humanist, secular humanist, Marxist—they all have evolution in their worldview. So I'd say there's primarily...there's probably like five major worldviews.

A quick search found Dr. David Noebel, founder and leader of Summit Ministries, of Manitou Springs, Colorado. Among other works, Noebel is the co-author of *Mind Siege* (LaHaye and Noebel 2000) with Tim LaHaye, famed evangelical writer of the best-selling *Left Behind* series of books. I asked Tyson how other cultures or ancient civilizations might fit into this schema:

> Oh...OK...I'm talking about the here and now...well...it is good to think about this. I think a major differentiation would be the naturalist worldview...of course that's evolution. You can either believe that everything is natural, or you can believe that there's a supernatural. That would be a big differentiation of how you're going to see how things go. And then of the supernatural, you can be a Muslim, or you can be a Hindu, and...but for some 'New-Agers' you kind of go along with evolution for...what is it called Gaia?... (*looking toward me for an answer*) what's it called when the world's like, alive?... ultimately it would come down to.. I think the first criteria would be...is everything natural or is it supernatural...and from there you're going to stem off into more (*trailing away*) ...and for the supernatural you're going to stem off into what kind of supernatural...what kinds of gods there are. Are there many gods, or is there one God. OK then...is it Allah or is it the Judeo-Christian God? If it's the Judeo-Christian God, is Christ Lord, or is he just the God of the Old Testament like the Torah?

Tyson continued on with this reasoning for quite some time. It is clear that for Tyson, whereas he may have earlier recognized change around him in the natural world, religious conviction and tradition could not be affected by the contingencies of history, human thought, or social movement. After running through a discussion of the various Creationist tropes against evolution, he again turned to a polemical view of science and worldview as he considered the validity of the science behind the formation of the Grand Canyon:

> So you've got two interpretations of the same data...and those interpretations are based on what a person believes. I believe a worldwide flood did happen... and the other person

believes a worldwide flood did not happen to the earth…so they both can sort of make sense of the data…but one of them has to use a lot more work. I think the evolutionary worldview has to…I think it could make sense of the fossil record, but it's so…you have to…it just seems much less realistic.

In my earlier read on Tyson's experience, I was taken back a bit by the clarity with which he worked with the concept of worldview but concluded that he would "probably be too lazy to change his worldview." As he detailed, the perceived potential loss of friends and family connections was reason enough to back off of some forms of self-interrogation. As a phenomenal experience, this was reason enough. In this second narrative, Tyson in his own way self-checks and confirms his story by seeing the ease of the Creationist narrative versus the more complex and harder work of science. A cynical take on Tyson's reasoning might even grant him an elegant application of Ockham's razor toward its practical reason (Bourdieu 1998). Tyson, as interesting as his case was, was clearly not *quite* on the precipice.

Another side of a worldview shift was represented by James. One of the atheist/agnostic students—he had made such a realignment of view long before I met him. As he detailed in our interview, the process of remaking a worldview was neither pleasant, free of affect, nor even completely peaceful now in his adult life. Beginning with what he described as turning him from religion:

When I was really young, I would readily go to church with my grandmother who's Pentecostal. I don't know if you've ever experienced Pentecostal churches…the shaking, and the speaking in tongues…stuff like that was disturbing to me as a child. It was frightening…it was such an intense experience for a child that age…it was just like sensory overload…they were just short of drinking arsenic and taking up snakes. It was scary to see grown men jumping across pews and knocking ceiling tiles out…my aunts flopping on the floor. Yeah, it was very intense, but…it was frightening to me as a child, and I would be reluctant to go.

As James shared with me, he no longer practices this faith. He described his current belief system: "Every human has this unique consciousness that is them. We're all connected in a certain way, and that energy has to go into something, go somewhere when a person is deceased…that unified force, maybe the universe is perhaps God." Describing what sounded like the God of Spinoza or Einstein, James was a long way from the old-time religion of his youth. As he continued, his recent adult life includes differing kinds of discussions regarding evolution that pull him in differing directions: "I actually constantly continue to speak of evolution and scientific matters with some of my best friends because they are higher educated people—like there's engineers, a nurse, international business, software engineers… one's a clinical counselor." Living at home and returning to college as a nontraditional age student James speaks about how his interest in science is received at home. "My father and I don't really talk about it. I live with my father currently, so I could afford to go here." When James reflects upon how he interacts with the rest of his family about science and religion tension is evident:

My grandparents know that I'm not as religious as they would like me to be, and my mother knows more about my views because I'm more open with her. She does not like the fact that I do not believe in Jesus.

As I asked him if this affects his relationship, he summarized: "Yeah, because I—it's hard to go around them constantly giving a lecture about 'you should be saved, you should be saved. You should take Jesus.' I'm like, 'I don't believe in him.'" This tone of slight animosity, while not indicative of James's total relationships with his extended family, does appear to grate on him. Typical of the kind of disconnect between the religious exclusivity of his family and James' growing views, he explained in our interview:

> I would watch scientific programs and things like that from an early age. I spent a lot of time with my grandparents and if I would say something about evolution, and my grandfather would say, *oh, they've got you too*. My grandfather is very anti-evolutionary theory, so, anytime that I would bring it up, or I'd be watching a program about it, it was kind of like - *that's evil, that's the work of the Devil, you know?*

As James explained, (*mimicking his grandfather's comments*) upon seeing him watching Nova on PBS: "The secularists…brought you away from the teachings of Christ." Unlike Tyson who sensed the edge of his worldview, and James who had experienced a gestalt switch in his, almost all the students I had the opportunity to speak with expressed no such dynamics in their lives, except three.

Along with Tyson, James, and each the preceding student narratives, each except one (Renee) had what I would consider to be a stable worldview. There were no major paradigmatic shifts in their thinking toward science or religion really at the forefront of their lives as they described them to me. So how does evolution fair when worldview actually changes? What conditions actually prompt a worldview change so that the significant narrative of science and religion in one's life becomes fluid? Does this understanding reorganize around a new identity?

Public intellectuals who advocate for evolution such as Richard Dawkins have a story regarding science and religion that is plain enough. Understanding evolution crippled and dismissed any commitment (albeit a fairly pedestrian Anglican one) to religion for him. Other notable atheists such as Michael Shermer have described a more extreme move from religious fundamentalism to his now 'skeptical' position as he calls it. My own experience was somewhere between the two. So, I knew in some sense, what worldview change might look like. Of the students I interviewed, there were a few whose relationship toward evolution was changing. The reasons for this are not perhaps the most intuitive, but real and important nonetheless. Given the prior tensions that the students so far have described, I did not expect evolution to be a catalyst toward worldview change and for the most part, I did not find that. What I did find though is perhaps more compelling for scientists and science educators who scratch their heads at students who shrug or dismiss scientific "truth." The following are the stories of three students who, almost certainly as I write now, are in the continued process of worldview change. One shifting from Christian exclusivism to agnosticism, one from atheism toward Christian exclusivism, and one who wishes to be happily Both/and but is being pulled strongly in competing ways by those in his life. Why does this happen?

5.1.1 Cindy

When Cindy and I spoke, she had settled on being an elementary education major, one of those who for the interests of science educators have the potential to lay the foundation of evolution education during their career. Cindy had come to Mason-Dixon as a transfer student from another regional college due to circumstances that she will explain. When her biology professor, Dr. Fleischman introduced me as a researcher who would be sitting in on the classes that semester, Cindy was one of a few students who said hello and seemed curious as to what I would be doing watching their class.

Cindy came from a small city in a mountainous region a half day's drive from Mason-Dixon. As she describes it, evolution was not exactly "taught" in the high school she attended. "I remember it in my sophomore year. I just remember it was my sophomore year." As she summarized the teacher's treatment of evolution, "she actually came out and said, '*well here's the chapter of evolution*'—and we watched a short clip on it, then she was like, '*well...we covered it!*' and then went on":

> [The teacher] would talk about that we were gonna talk about evolution. And actually, some of the kids didn't come to class the next day because they didn't want anything to do with it. There were about five students out of the thirty in my class and I think we had almost 30 in the whole grade that didn't come. The fact that the students were physically rebelling...which I think is stupid...but they just weren't coming to school! A couple of them actually went to my church, and they just said that it wasn't a part of their belief system, and they didn't feel that they should be forced to learn that. The teachers just didn't want to go too in depth with it because they were afraid that they were hitting points that the students didn't want to hear due to religion. There were a couple of students that went to the principal whenever we started talking about evolution, and they said, "We don't think that this should be happening or going on." And you know, with our school board being the way it is probably would've said, "Can't talk about evolution!" We actually, during our graduation—you know you're not supposed to have a preacher come in or do a prayer or have—oh, yeah, we had all that. The preacher from the church across the street came over, gave a prayer, all that stuff, so it kind of neglected any issue that you're not supposed to involve church and state.

As she explained, her discussions with her teacher both elucidated that she "wasn't religious," and that "the administration has come to her and asked her why she wasn't presenting [evolution] as wrong." The teacher has since left the teaching field and now is a pharmaceutical representative.

Cindy continued to explain the anti-evolution discourses of her social world at home. The events of the classroom apparently had a reflexive relationship to some degree with the discourse of her church youth group:

> Yeah, we discussed it in youth group. Usually it rolled into an argument because our youth director would—he would mention it and he doesn't agree with it, but he's saying maybe those aspects pull, you know, into religion or whatever. And other times, actually a couple of people that left the bio class, they were there, and they were like, 'We don't want to talk about this. What are we doing'? And then it would just kind of disintegrate and end. But really the ones that we fought the most about were really probably evolution and abortion. Because in that community word goes around...and you would be shunned.

Cindy also gave a telling glimpse into a possible interpretation of science generally within the community: "Our local libraries and stuff back home really didn't touch on science whenever we would go there, as far as magazines and that kind of stuff. They were kind of back in that corner that you never approach." Curious as to how this played out at home, I asked Cindy to explain whether evolution was discussed at home:

Absolutely not. My mom is…I mean she doesn't go to church a whole lot, but she's heavily into that thought that evolution is *not* the way that we came about. I mean, anytime that I've tried to approach that situation, even until this day if I call my parents and strike up a conversation it'll be like…you know. (*gesturing with what appeared to be her attempt at aping her mother's stink-eye*) I did ask my mom one time, 'What do you think about evolution'? And she just told me, 'Don't want to talk about it. Can't believe you're even thinking about it'. I mean well—she just told me that if I get into that, and believe in that, that I'm pretty much going to hell—to put in bluntly.

The clarity of this anti-evolution discourse within her life-world belied something Cindy had mentioned early on in her interview regarding evolution and her religion. In each of my interviews, I had started by prompting the students with a minimalist question by which I hoped to elucidate as raw a response as possible. Through this response, Cindy left subtextual strings which I could easily pull. Telling me what evolution means to her, she explained:

I feel that it's true. I mean, you know, growing up in my community, I was taught to not believe that at all, but I mean I can see the science behind it. So I guess that I'm just kind of questioning it all.

The bread crumbs that Cindy had dropped throughout her narrative told of a more recent move to a more orthodox view of scientific data. As she said, "evolution didn't really interest me in high school because I really didn't think about it." So what changed in her life? Had she, through reading just the right amount or type of scientific description, seen the light? Knowing that she had been framing most of her community in a negative light, I approached another possible perspective of worldview and change—her religious faith. Had it changed at all during her life? "Well, it definitely has." I asked her to detail in what way:

I started going to my youth group when I was in the second grade. Whenever I got baptized, I kind of—the more I think about it now, the more I think I just got baptized kind of because everyone my age was going ahead and getting baptized. I wish I would've kind of had more time to think about that. I've rethought whether Christianity should be the ultimate religion as taught in my community. I've questioned whether the Christian God is the all-powerful God, or if there is a God. That's really…that's pretty much the big one.

What prompted this reflexive dialogue? Reflecting regretfully on the normative function of her baptism, Cindy turned to the present:

Now I have not been to church in about three years. There were some things in my life that happened… and the congregation started talking about it. After that I kept thinking, 'Well, is the church really here to go to church, or are they here to gossip'? So I really haven't been back to church. Plus, you know, with me questioning everything and how everything evolved, I just don't feel like church is really the right place for me at this time in my life.

Naturally, my interest was piqued. What could the congregation be gossiping about? Within the content of her narrative, Cindy linked "things" happening, the social discourse of the church, and now a questioning attitude which included the possibility of biological evolution having taken place. With only semi-structured questions in front of me, I stepped forward into her life. What had happened?

Okay…I mean…(*cautiously, and then resolutely*) …I'll flat out tell you. My freshman year of college…I got pregnant. I got pregnant. You know the guy treated me very, very wrongly. As soon as I got pregnant, you know, the church was like, 'Oh, my God. Why would you do that? So much potential', and they thought that my boyfriend had corrupted me.

(*Pauses, takes a deep breath, and looks away briefly…reestablishes eye contact*)

And I did get an abortion. However, they do not know that. I…you know…I'm not going be out going out and saying, "Hey, I got an abortion!" It's not something I'm proud of. It's just there was a lot of chance that there could've been something wrong with the baby anyway because I had had some nice little drinking binges before I even knew. So that was kind of the consideration I was going at.

Again, like the surprise of speaking with Julie about linking evolution with death, Cindy threw me off my interview game. Finding a way out of this tension without diminishing the impact or importance of her story, we continued:

There were a couple of families within the church that I still talk to. They were very supportive of me. But for the most part, the church people would just talk behind your back, and things would make their way around the neighborhood. Our neighborhood is extremely small. Within that community, it seems like everything gets around in about five minutes. Gossip wise at the church, I think it was just a big assumption thing, you can tell whenever you go back and they look at you, you just know what they're thinking. And I heard a couple of elders one day in front of the church talking about how they don't really know if I had an abortion, or if I miscarried, or—you know. The church was very much so not supportive of me as a whole and I just—I couldn't go there anymore. I couldn't face it.

Like the constrictive social networks that Wuthnow (2005) sees typical of exclusivist communities, Cindy's experience is fraught with gossip and insularity. Curious as to how this experience has at all changed her religious identity or practices, Cindy explained. "I don't know if there is a religion for thinking that there's a higher power, but not knowing what it is. But if there is, that's kind of how I see myself right now." I asked her how this developed—first at South-Mountain State where she first went to college, to her life at Mason-Dixon:

At South-Mountain State I still did, because South-Mountain is close to home, it's like an hour away. There's a place called Spirit-way that was supposed to be kind of like you just come in casually, sit around, talk about things. I thought about going there, but honestly, no. And since I've been up here at Mason-Dixon, I haven't really thought about it. I have so much going on…work and school.

The social structures of Cindy's recent development began to disclose themselves. As the role of the church in her daily life has faded, she both has family that remains committed to a fading world for her, while moving through higher education has opened up new ways of being.

Cindy has two older sisters. The first, similar in her attainment of college education, and in this case a graduate degree, would talk with her as what sounded like the most meaningful confidant in her early adult development. When I clarified any dialogue regarding evolution more recently she detailed:

> She kind of has the same kind of feelings as me. Like we talk a lot on the phone about—she *does* consider herself to be Christian. She does go to the Spirit-way place where it's very casual, but with her you can talk about it because she's been in it. The rest of the family? Absolutely not. They don't want to talk about evolution.

This sister, who had trained and works as a nurse practitioner, could likely not have avoided evolution in her scientific training. When I asked about a typical conversation that she might have with this sister:

> It was around the time after the pregnancy. I was having a lot of questioning and couldn't talk about it to mom, and I brought it up to her. I was like, you know, 'Is it bad for me to say that I'm questioning my religion'? And she said, 'Absolutely not'. She's like, 'You're in college. People in college experience new things. You're not in Eastland any more. I think a lot like you now'. Seeing how her husband's father had just passed away...so she had that mourning time of questioning and, you know. Pretty much we just talked about our thoughts on: is there a God? And how do we evolve? And we both came up with the answer that we're still questioning.

Let us not forget that this outlet is but one of a larger family structure, which effects dialogue, and in some ways intellectual development. Her mother and community members that she detailed were resolutely against evolution. And when I clarified about her other sibling, she typified her other older sister's attitude: "She thinks very much that God is superior. She'll say that God is the reason that everything is happening all the time. I mean, *you know*...." I asked her what this sister does for a living. "She is a special education teacher."

Turning toward the world of Mason-Dixon, Cindy had already described not thinking of church or active organized faith practice due to the rigors of college work/life scheduling. Important to remember in the political economy of higher learning is that most typical college students (such as those of Mason-Dixon) must work to makes ends meet. Curious as to how discussions of evolution might play out within the structure of her day-to-day life, I asked her if she could recall any discussions she had regarding evolution outside of class:

> My best friend's boyfriend is atheist, and he has had heated discussions whenever I'm in the room, not necessarily with me but....he goes to another college but he's around a lot, though. He will bring it up, and usually I'm with my friend, and she just tells him that he's an idiot and stuff, and he gets heated up into it. So I mean, not really an educational conversation on it, but I've been around it.

The discursive speech of actors populating Cindy's current world is made more complex by the odd spatiality, abbreviated temporality, and saturation of current social networking communications. Wishing to untangle what I was hearing as evolution being equated with atheism, I asked Cindy to clarify. "Okay...Jessica and Shelby, those are my two really good friends—this actually involved Jessica and Shelby's boyfriend":

> Because they would both call me as they were arguing being like, what do you think? What do you think? I'm like, okay, I'm in the middle. Okay. Mark is an atheist. Jessica grew up

Christian, and she's my best friend – but they're both very hard headed. They don't care to tell each other what they're thinking, and the heated debate was atheism.

I asked her to set up the reason for debate for me:

Mark finally posted his religious views on Facebook, and it said Atheist. And that is forwarded to everybody. The Jessica came into it, and Shelby was kinda in the same situation as me because my fiancée is not atheist, but he certainly is questioning. So me and her would take up for each other because we were both raised in the same community. So it's kind of hard for us to be with people that kind of were not on the grain, they didn't go with the grain of our community.

(The following is stated in one run-on breath)

But Jessica just absolutely could not believe that Shelby was with an atheist... and she called Mark and was like...how can you be atheist?...how can you live here, how can you celebrate Christmas if you're an atheist?...and then Mark would bring in the facts, oh 'that Christmas wasn't actually in Christian religion', it was brought in by I forget who, he said the pagans or something...but and that has really been a big issue because they just finally agreed to disagree...and then Jessica would still talk to me about it and Shelby would find out from her that she was talking to me; it's just this nice little triangle of confusion.

To clarify, Jessica, who had gone to the same high school as Cindy, wasn't apparently as openly supportive of evolution. "I've asked her what she feels about it and she thinks the same as anyone else in my hometown would say." Cindy also discussed another friend unassociated with the social world of home whom she met through her fiancé, both of whom are art majors, whom she described gravitating toward. "She's really open about religious views, and anything....that is why I gravitated to her."

Closer to the immediacy of intimate relations and the possibilities of dialogue they prompt, when I asked a question regarding the teaching of evolution in schools, Cindy discusses with me her fiancé's background and attitude toward both evolution, science, and religion:

I am engaged and he has conflicting feelings too. I mean, I would say it's okay. I mean, we can't really raise a child to be what we want them to be because like for me, for instance, I said I grew up in a Christian environment. But whenever I got to college, I started questioning that anyway. And probably the biggest person that I talk to is my fiancé. He doesn't really—I think that he believes that there is a higher being, but he doesn't know exactly what it is, and so I guess I was always raised to be with a Christian guy, but I love him, so I'm not going to be with anybody else. But he really makes me think. Because he will ask me questions like, how do you know that your god is God? So I think it's better to start questioning stuff whenever you're younger.

When you talk about you're being engaged and your fiancé is conflicted about that too, can you tell me about the conflict? What's going on? What is he conflicted about?

He's conflicted about just why we are—why we're here and he doesn't really—he believes in some sort of higher power, but he doesn't know whether that's a God or what that is. But I mean he, you know, he feels that something had to bring us about, but I think that he would say we're more of the scientific view of it. He's been absolutely everywhere, though, and I think that that might be why he kind of just—he's seen various parts of the country, and different cultures, and different types of people. And he's just not convinced that one vast religion should be able to say this is why we are.

Seeing that Cindy was simultaneously negotiating many different discourses regarding both science and religion, I asked her to clarify how, when, and where she does this:

> I do this partially in my head. But probably my first encounter was when me and my fiancé had a roommate, Scott who was Jewish. He was the first encounter I really had with someone that had been hardcore in another religion. And he got to talking about it, and it has a lot of similarities to Christianity. And another person would be my sister, my oldest sister because like I said she's a nurse and so she's been exposed to a lot of science, so she questions a lot too. We kind of balance back and forth because she knows that I really can't talk to the other members of my family about that.

And when I asked Cindy to reflect on whether she now feels conflicted about evolution:

> Yeah, whenever I was in church it conflicted because in Christianity, they don't think evolution is the reason why we're here. So obviously it conflicted then. I mean, it still conflicts now but I'm just not subjugated into my church anymore.

And how this plays against any feelings she has toward her not participating in faith practice right now:

> From time to time, whenever I'm not busy, I think about it. And I don't really like that I don't have any kind of religious thought right now. I was so used to going to church and youth group and stuff, but at the same time, maybe this is pertinent, or not pertinent, I think a big reason that I have not picked a church, or attempted to go to church, is I have not wanted to have that discussion with my significant other. Oh, he's already told me that I could go, he's like…he doesn't want me coming back home and talking about it because he don't want to talk about it, so…(*pause*) and he's my only person up here.

I asked Cindy to qualify a difference between herself now and the young woman in high school. I reminded her of her earlier statement about not considering evolution in high school, and what did she attribute to the difference between them?

> I think it's mostly about being scared. I think in my teen years is when I really realized that death was factual, so I think just I didn't want to think that there wasn't a heaven, or…you know. I mean, I guess I feel that if evolution did happen the way that we we're talking, being strictly scientific, I guess what was what made me scared because then the whole myth of God or whatever, wouldn't be true. And it would just be like—say a dog dying—dog dies, she's gone. And I would just be gone. So I just don't think I wanted to think about that back then.

Like Julie before, but perhaps in a bit less dramatic fashion, Cindy has ended by equating considerations of evolution, deep time, and the existential quality of her life.

It is likely that this discourse, the disconnection from the "subjugation" of her home church to an identity of "believe there's a higher power, but not knowing what it is" might encourage those committed to traditional liberal education while horrifying some evangelicals. What is clear within this is that this move itself opened a space by which evolution, along with experiencing other people, worldviews, and cultures slipped into and grabbed hold as a reasonable, if at least possible, foundational discourse. In Cindy's case, evolution had not changed her religion per se, but the change of her religion allowed evolution to come into focus as a reasonable possibility. The structure of Cindy's life—the social networks she has established practices within, the economic reality of her busy college life, and the deeply important affective domain of her being, both with her fiancé and increasingly disconnected from the shunning of the home and church, each were at least structuring new possibilities within her worldview, and possibly foreclosing others.

5.1.2 Renee

Renee was one of the first students with whom I spoke. A biology major with sights set on a graduate school pharmacology program, I came to see her as one of the more bookish students I interviewed—at least based on the amount of hours I would see her reading her texts quietly in the rear of the student union. For my interests, Renee's story began immediately with a hook. Asking her about what evolution means to her, and if her views have changed over time:

> That's a good topic because I am a new Christian and also a biology major. I think that creation and evolution can definitely go hand-in-hand. I've always been science-minded and the more I learn both in church and here, the more it seems like they're not mutually exclusive. So I do believe in it. I've always been kind of aware of the whole origin, origin event, the natural descent. It never seemed to quite make sense, but you can definitely see how species can—I call it adapt rather that evolve—adapt to their environment and change over a period of time. It's the whole Big Bang Theory that doesn't quite seem to make enough sense to me. That's where I have the issue.

Yet another student had linked evolution with the larger cosmological phenomenon of the Big Bang, and ascribed this as negative. Although I would certainly investigate this further, I made note and stuck to my interview protocol. How did she feel about evolution?

> Like I said, it's easy. It's easy for people to think things just happen over a course of millions and millions of years. But it gives people no hope in the end. If we evolved from molecules and now we grow, grow, grow, and die? What is after that, just nothing? There's no concept of a soul and that just doesn't seem to make sense to me. There's got to be something more. We need to answer for the things we do over the course of our life.

Renee went on to detail what by my judgment was a fairly robust preparation in both biology and chemistry. As she described the high school offerings, I also asked her to detail her school experience with evolution:

> Oh I loved [science]…after the general biology course, from the electives there was a field biology class that I took, a marine biology I took, and an anatomy and physiology. I even took an extra chemistry course. And my field biology teacher, especially, he loved what he did. He just really inspired that in the students. High school got more into the evolution side just because I was doing more, I guess, intense classes for high school like the marine biology and anatomy. They got more in depth about it but it was always—that it was the only way. No one ever said, 'Well, some of you might believe this'. That wasn't even part of it. Like this is what happened. This is how man started as little monkey to the ape, to the…I mean, it was just…that was facts. That's how you're taught…you *need* to learn it.

As Renee continued, she vaguely remembered the influence of high school peers who were church goers who tried to give her their opinions on this. Given her new-found identity as a Christian, recalling this memory seemed to her a bit odd. We turned to the influence science museums had on her early life:

Yeah. I used to love dinosaurs. When I was little we used to go to museums. And actually my dad took me to the Creation Museum a couple of months ago. I just went to the Bodies exhibit at the New Providence Museum. I loved it because, like I said, I love the way—just the way the body works, how it reacts to things. I mean, it was fascinating. I could have spent three days in there. We also went to the Natural History Museum. This was during that period where I was going to church and kind of learning, so there was a lot of confusion there about all the little plaques in front of everything says, 'This animal evolved six million years ago'. And so that kind of threw me for a loop.

Renee relayed the influence of her father's dialogue on these trips:

Well, he very, very, very strongly believes in creation. And I was kind of curious, just wanted to see how they presented things because all my life it's been biology, evolution and that kind of stuff. And it was interesting. It really made me think.

Given Renee's recent trip to and her father's advocacy for the Creation Museum, I was curious how this experience shaped her views:

Well, there's little model T-rexes with flat teeth because they were herbivores back then. And that kind of was – because all the fossils we have are sharp teeth. And I said, 'Well, why aren't there fossils of these things with flat teeth? You know they were carnivores'...'Well, in the great flood, they were all destroyed. And after the flood, there were herbivores—or carnivores and...'. Yeah, well, I also need to explain that he gets...like I said, he's really... he believes in it *very* strongly.

My curiosity at the before and after narrative of Renee's religious epiphany was now overwhelming. Seemingly so important to setting the ideological tone to her interpretation, I asked the obvious. What happened that she became a Christian?

I don't know. I started going in January and just learning the other—the flip side of the evolution—just learning about God and now it's just everything here. I mean, I spent the holidays in the mountains in Gatlinburg and it was so gorgeous. And I'm like, 'How can this not be the work of a creator? It's beautiful'.

Although her conversion story was pleasant enough, I wanted to check whether any specifics perhaps nudged things along. I asked her whether a person in her life aided this. "No, I think I sought it out on my own. I said my dad's been going for years and years and years. And I just decided it was time. It just felt right." A critical bit pops out when I asked whether Renee had experienced any major life upheaval recently:

I got divorced this summer. My divorce was finalized in June. And I still didn't start going until January. But I've been blessed a lot with family and loved ones. And good things have happened, have come from it. And I'm just like wow, I feel like someone's really taking care of me.

Renee went on at length detailing her experience. Interestingly, the circumstances of her divorce, which centered on marital infidelity, mimicked her father's:

It came into his life when he got divorced. My mom cheated on him. Looking back now, he was beyond rock bottom. I don't know how he made it from day to day. And for some reason, he was brought up religious when he was a kid, but he didn't go to church all his life. And for some reason, he was led back there. And now, I know from experience, that when you have trouble, I mean, these people just came out of the woodwork to help me and be with me. And I think that's what brought him in more than anything else. It was just a support group that was automatically there. They didn't judge him.

As she shared, a friend from work invited her father to an adult Bible study group at a Church of Christ that he attends. The influence of a church in her life must have been a bit jarring, given the contrast of her mother's influence, who met Renee's father while he was stationed abroad in the military. Renee shared her mother's relationship toward science and religion:

As for her science education, it probably was minimal, at least as far as she can remember. So she is interested in it, would be interested in learning more, I think. But as far as religion goes, she's very anti-religion. I feel it's an attack on me, which is probably odd because you know, being my mother, you think she would try to at least be supportive. But I remember when I got baptized I was worried about telling her because I didn't want an argument. My grandfather was in the hospital. We thought he was going to die. And I was talking about, you know, I wish he was right with God. And that's when she said, 'You realize we are atheists'. You know she said, 'I know you're not', but why would you bring that up right now? I bristle when I think about talking about religion with her.

We briefly discussed the difference in religious history in the USA and Britain (Renee's mother is British). This turned back toward a discussion of this influence on her own spiritual growth and science education. I asked her how she balances the influences of her science education and her religious education. "I've been asking a lot of questions." I asked her to be more specific:

Mostly my dad. He's a biblical scholar nowadays. He's very, very, very much a Creationist. But not really a whole lot of experience in the sciences. He was arguing about carbon dating, but he didn't really know what carbon dating was, so he couldn't kind of explain it. He doesn't really have the scientific background. I think he's very one-sided. I was sitting, doing homework at my dad's house, and it's biology. He's very opinionated. He likes to tell you exactly what he thinks. And that's when he first started saying, 'Well, you know that's not true. I can't—this is just ridiculous that they believe this. Can you believe this? Do you think you originated from this'? And I was like, 'What'? And then you start talking to your other high school peers and kind of seeing the other side of things.

Recalling Chap. 4 where Renee was "shut down" reaching out to another student, I was curious as to how she planned on negotiating the social space of science. Asking her if she thought her new Christian identity proved a challenge in this:

Yeah…it's a struggle. Biology has always been my passion. I've always been into science and I think I always will be. Religion has kind of thrown me a little off kilter. So I'm just trying to balance that and one of the things is, I love *National Geographic*. I read it all the time. It's a very evolution based magazine. They do try to report, I think, from a neutral angle, but it's not. It's very slanted. So then I also ordered another magazine called *Answers in Genesis*, which is a very Christian side of things. I love reading things from both points of view. You know, I'd love to read *one* article from a Christian in a evolution standpoint, just to see how everything could change so easily. In class month it was Darwin month and it's—they don't give you the option—they don't present both sides of the story and let you pick. It's only evolution, that's it.

Curious as to how she framed "both sides of the story" I asked her to explain:

Basically well, in last semester's Mock Trial, like it was evolution versus the teaching of creation. But back then it was backwards (referring to Scopes). He wanted to teach evolution.

And I just wish they would present both sides of the story—like, this is what happened, Adam and Eve and that kind of stuff. I wish they would at least just even touch on it. A chapter would be fine. But they don't present it at all…it's basically taught as fact. But there is a guy—a pharmacologist for the last 15 years, who is teaching a creation versus evolution course at my church. I went and met with him before it started. And I said, 'I'm really hoping you can help give me some balance here'.

The pharmacologist in question works for a Fortune 50 company, has published extensively in the chemical engineering literature, and is an active Creationist evangelist.

I asked Renee to reflect on the entirety of our discussions and sum up her standpoint regarding evolutionary theory, given what we had discussed. Through a halting explanation, she got out that genetic mutations do happen, and that natural selection works—but she prefers to "call it adaptation." Her hesitancy, no doubt prompted by the new and competing Creationist narrative of her church, is made explicit when I asked her to explain what's lacking regarding evolutionary theory:

That's a hard one. Like I said, we know it happens. We know species do change over a period of time. But it's more of, again, starting with the molecules to man theory and over millions and millions and millions of years. And as we grow through our lifetime, it happens over a much shorter period of time. So I don't see why it's happening in a short amount of time here. But in the past, it's taken millions of years to change it. I think we just don't know enough about it. I think that's the main problem. But you could apply the same to Creationism. We just don't know.

"Molecules to man" is a common metaphor in contemporary Creationist critiques of evolution. As a sort of semantic shell game, it gives the impression that Creationists accept a good bit of evolution, while rejecting "macroevolution"—change that violates the immutable concept of "kinds" from their literalist interpretation of the Bible.

Renee and I discussed whether she thinks schools should actually teach evolution. "Well, we know [teaching evolution] happens, so you just kind of have to suck it up and know it front and back for the tests." Sensing a bit of disappointment from Renee regarding the state of creation in education for her, I asked her if she could change our education system's treatment of evolution, how and why would she, if at all:

Yeah, I absolutely would. Through elementary school and middle school, I would present it equally—like this is the story of creation and this is the story of evolution. Because I didn't hear anything about creation really until I decided to seek it out on my own. And then once you're in high school you can kind of develop your own ideas, I think we should still maybe present a chapter on it or mention if. If you believe in creation, you can think of it this way. If you believe in the evolution, this is how it originated. Right now for me, it's more me about learning. Like, I'm going to listen to everything the guy says at church and kind of file it away. I'm not formulating opinions yet. But like, they've got the little models of the dinosaurs with the flat teeth because they were herbivores. That is just very interesting to me. So I'm going to file it away, and I would love to bring it up like in a biology class at some point. I'm also kind of afraid of being attacked when bringing up creation.

The Answers in Genesis Creation Museum presents carnivorous dinosaurs as being herbivorous prior to the "fall." From a diorama placard titled *What did dinosaurs eat?* the Creationist spin on animal habits: "Before man's fall, animals were vegetarians. All the beasts of the earth, not just the 'beasts of the field' that God brought to Adam to name, ate only plants."

Counter to the dissolving commitment toward religion that Cindy was experiencing, Renee was newly baptized as an adult into a church whose vocabulary was steeped in Creationist concepts, and saw *Truth* as totally and inflexibly emanating from just one interpretive frame, which was as they saw it, the right one. For both Cindy and Renee, the circumstances of their social lives were scripting a small set of the wide range of possible discourses they might have regarding science and religion. In both cases, the limited repertoires that were closest to their homes and lives had prime impact on how they were coming to see both science and religious faith. In neither case were more moderate inclusivist theological dialogues proximal to their day-to-day lives.

5.1.3   Nolan

Prior to his volunteering to take part in an interview, I came to know Nolan for his affable personality. Often seen laughing and chatting with others before or after class, Nolan appeared well-socialized into the culture of the biology students. A pre-med student, he described his interest in science. "Well in general, biology has always been something I was interested in. I took regular biology and AP biology in high school. And particularly, I like the genetics and the cellular kind of stuff."

As Nolan remembered, evolution was covered in a unit within his biology class, in which he remembers "taking about a page of notes." As he explained, the social controversy over evolution in the curriculum was also covered in his school. "I think it was either eighth or ninth grade, in history class. I think they kind of went over it." As he explained, discussing the history of evolution in America also prompted some slight discord. "It brought up a little bit of a…brought up some questions among some of the students in our class. We had some people going this way, and one the other." As he remembers, only the Scopes Trial and the associated discussion of the Kitzmiller Intelligent Design Trial prompted this. I asked Nolan to situate himself in these discussions:

> I feel like it is an appropriate subject to be taught in schools, at least. And personally, I feel there are some things that definitely make sense. I'm kind of open to all of it. I think it's all very interesting and that's why I'm a biology major.

Running right down the middle with a Both/and ontology, I asked Nolan for more detail regarding his understanding of the science: "The things that do make sense, for instance, like how everything they say started out as prokaryotic organisms, and they were photosynthesizing and made the oxygen for atmosphere to make—to allow larger beings that respirated." As he continued, he cited fossils which show homologies and evolutionary relationships across time. Turning toward anything he felt lacking in evolutionary theory: "How supposedly we came into being through just oxygen and energy and things as that, and then also…(*trailing off*)…like, just how similar all of—oh I don't know. Like…it's hard to explain." Nolan turned his answer back to what he does know. "But like I said, we're all—even us, including animals and everything, we're very similar. It's very possible that we all came from one specific common form, at one time, and—I don't know."

I turned Nolan back toward those in his high school class that had issue with evolution. Asking him to imagine what their problem was with evolution: "For people who are *very* into their religion and believe in creation—that a higher being was what was behind us being here, and not humans coming from, you know, a lesser being." Nolan concluded. "I guess mainly what most people would have a problem with would be a human evolving from some lower level species." Although I need not ask more about Creationist complaints toward evolution, Nolan's construction of the "*very* religious people" had my interest. Asking him to explain:

> Well, those are the kind of people who, like, if they—well, for one, they attend church very often. They're regular churchgoers, and they don't believe in any evolutionary type things being taught in schools, and they will normally defend themselves about something in front of you, and just won't let you let someone say their opinion about something. They normally are there to combat or defend their religion from what I've experienced.

As Nolan had made clear, he regularly attended a Southern Baptist Church during the early years of his life. Given the tension that he saw evolution causing in some of the educational discourses that he had experienced, I asked Nolan to walk me though his faith development since childhood. "When I was younger, I was forced to go to church. I had no other choice. Now it's more of when I do go, it's a more special occasion type of thing, like the holidays, and stuff like that." Echoing parts of Cindy's story, his developing religious identity has sputtered while in college. "Honestly, it's just like I have hardly any time for that with school, and work, and sports." His discussion turned right back toward a critique of the *very* religious. "Some people are very into their religion and they believe that they're right. Nobody else can have other beliefs and that to teach that they can is unquestionably wrong. Since [being in college], now that I've kind of had free will, I've been able to—I just don't go as much." Nolan, both clearly felt some kind of sting from his heritage as a religious exclusivist, while clearly wanting to have Both/and. "It's probably not as big of a thing in my life as it used to be, but, I mean, but I still consider it something somewhat…a somewhat important part of my life."

To my surprise, although his practices have moved on, parts of his conceptual relations toward being Both/and hadn't quite gelled. I asked him whether his religion conflicts with evolution. "Well, I would say evolution definitely conflicts with the religion. Like it's the exact opposite of the religion." Asking him to explain this definite conflict, how does he balance the influences from his faith and from science?

> Well, I just kind of take it as it comes. I pay attention, remember this stuff so I do well on the test. That's the most important thing to me right now, so that's what's I kind of go for, and when I do go to church, I listen, pay attention, trying to get the morals and values type stuff from it, and not so much tried to look to find something to disprove one thing or another.

Having been raised in the religious exclusivist tradition, Nolan now simply wanted to be Both/and. So far, my most illustrative question in these regards was whether students would have issue with Creationism being taught alongside orthodox science in public school classes. "I, personally, would have no problem with it whatsoever. I would feel more comfortable with both sides of it being presented in classes." Nolan was simultaneously charming and vexing in his go-along-to-get-alongness, as the glaring inconsistencies he worked with began to amplify. This came more to the fore as his narrative turned toward a local flashpoint for evolution and science. Nolan, in speaking about a conversation with his stepfather and beginning to unpack the dialectical structures of his life, shared the conflict that ensued from his planned trip to the Creation Museum:

> I was going to the Creation Museum and my stepfather goes, like, 'Oh, you're going to that thing? You're actually going to walk through there'? I was like, 'Yeah'. I was like, 'I'd like to see both sides of the arguments'. And he just said there's no way. How does this one guy create the entire...and that's about pretty much the extent of the conversation.

As Nolan contextualized his stepfather's position: "My stepfather, I'm sure he does believe in evolution." As he explained, "Well, for one, he straight out told me that he believes that belief in a higher power is absolutely... that he thinks it's completely stupid and idiotic." Asking Nolan to detail his impressions of his stepfather's relationship with faith: "He'd never gone to church that I knew of, recently anyway, since I've known him, and he just...." Nolan's stepfather is a clinical psychologist. Nolan hedged forward. "He just doesn't come off as a very—I don't know. He seems like somebody that likes to see more hard evidence of things." Naturally, given Nolan's Both/and inclinations, I wondered about his impressions of the Museum's "science." As he tells his stepfather: "I told them that it represented both sides of the argument pretty thoroughly." Past thoroughness, I wanted his adjudication. "I found a lot of their—a lot of the stuff they are presenting pretty interesting, and I thought it was done fairly well." Nolan's sole qualm with the museum came from an exhibit displaying biological rafting, a concept adopted by the museum from the orthodoxy of science.

As Nolan continued oddly enough, the messages from Nolan's home front are not as predictably clear as one might imagine. His mother sees things quite differently. "She doesn't believe in evolution. She goes to church every Sunday, but she's kind of—she hadn't been going for a while. Now she's been pretty stern in going to church a lot."

Although the discourse regarding evolution in this half of his parents' homes was polarized, what of the other half? As Nolan explains to me during the discussion of family dialogue, his father and stepmother never discussed such matters. In the interview protocol I wrote, there were more than enough times to cue and recue a memory about this. Whereas evolution prompted a clear response from one side of his parents, the others remained silent. As Nolan commented toward the end of our last interview, he had grown curious of this and planned on asking.

Interestingly, there were also plenty of opportunities for Nolan's social milieu around campus to chime in regarding evolution. I opened up the world of his campus social life by asking if they ever discussed evolution. "Every once in a while. It depends on the friends I'm with." He continued: "If I'm discussing evolution with somebody who's fairly religious, they attend church every week, they're not gonna be—they're more hardnosed and they're not gonna be more open to talking about it." Friends countering this "would say that it's definitely a possible thing and that it does have some backup and support with what scientists have found out about past biology." When I asked Nolan to share an example of interactions between these two groups of his friends, Nolan explained:

> No, they've never discussed it. I've never seen them discuss it…(*Pauses, biting his lower lip, staring off in deep recall*) Actually, now I think about it, just the other day, my friend that I was just referring to, and then my girlfriend, did get into a little argument about it. Somebody had said something, just randomly, about God or the Bible or something. I think it was my girlfriend, and then my friend…he said that he completely denied any existence of God, or a higher power. After that, my girlfriend just kind of started yelling at him, to be honest.

Again, without my prompting, the discourse of Nolan's life, like Cindy and others, had moved from evolution to a dialogue about the presence or absence of God. As many of the Creationist students had done before, they invoked religion as against evolution and vice versa. Little theologically moderate dialogue seemed to crop up in these students' lives. After having recalled one incident, Nolan's memory for such conflict was more attuned. He described a scenario after a class where evolution is discussed:

> Anytime [my friend] mentions anything of the sort—just like I was talking about my girlfriend, as well. They were both together one day, and I mentioned something about it after class. And they got into a heated argument with me about it, when I was just telling them what happened in class. Dr. Wayne had made a comment about how one of his reasons was that it proves that evolution was a valid topic, and that it invalidates Creationism in some way. He had mentioned something from some verse in the Bible, and I told her that, and my friend that, and she got very upset with that.

I asked Nolan if he knows why this upset her so?

> I want to ask them that. I don't understand, because I'm the kind of person where I'll let you say what's on your mind, what you believe, and I'm not going to judge you for it or anything, but they're very quick to judge about that kind of stuff.

Subtlety lost on Nolan, my own interaction with Dr. Wayne might have foreshadowed as much. In my interviews and in class, Wayne was clear about his betting with Pascal. Dr. Wayne called on this move when the implications of evolution grade toward randomness or even worse nihilistic premises in lectures. This kind of comment might pass most ears by without nicking them, but to Creationists such openness to metaphysical uncertainty cuts directly at many students' conceptions of appropriate theology and ontology. Without prompting, each time the discussion tended toward Creationist complaints about any uncertainty evolution implied, he delivered what sounded like a trope he had sounded many times before. "Do you know Pascal's wager?" beaming as if he had solved one of the great mysteries of life. "I figure, there's nothing I have to lose, where if others are wrong,…well then…." Indicative of the nature of disagreement between exclusivist Christians and Both/and students and faculty, it is as if the classroom itself is the venue by which the power over what counts as appropriate Christian ontology is partially delineated. As schooling from K-12 and the university afforded, varied religious ontologies interact in ways that might otherwise be limited in public life.

Curious about the extent to which Nolan's religious friends rejected evolution, I asked him if there were any examples of friends who found a middle ground:

> Well, let's see. One of my friends who actually doesn't believe in any creation type of things...he even went to a Catholic school and everything up to high school. He's just the complete opposite...the other side. They—he didn't have any problem acknowledging [evolution] or accepting it.

This left me with one line of questioning regarding his social discourse around evolution. What was up with his girlfriend...why so angry?

> Well, my girlfriend is *very* into church, and she goes all the time. And any time I even bring up anything about evolution in biology class or whatever, she kind of just shoots it down. She says, (*mimicking curtness*) 'Oh, well, people who believe in evolution and don't believe in God, are going to go to hell'.

Given that Nolan was a biology major, and therefore evolution may be a topic of conversation more often for him, I wondered if this type of conflict was common:

> It's normally more of me randomly mentioning something from class, or that brings up something, and she'll just be like, '*that is so stupid*'. That doesn't make any sense, but then I'll be like, yeah, well, you're still going to sit through the class. I got to do it regardless so, and I normally just de-escalate. I don't try and like, try and push anything, or start anything...

As I mentioned at the onset of Nolan's story, I would often see him, affably talking and being collegial with his classmates around the department—almost without fail—his girlfriend very securely locked to his side. As I closed my last interview with Nolan, and as happened once before, a scan across the student union spotted Nolan's girlfriend sitting petulantly, scowling at me.

The structured practices within which lives had unfolded for students also delineated quite a bit of the possible avenues by which they walked and would continue to walk. This is not to say that individual agency was not possible. Rather, this is to acknowledge how culture has already shaped the types of practices which we find natural, comfortable, and reasonable. Cindy and Renee were seemingly passing each other as they switched ontological positions. Nolan felt the polarization of the discourses of his life but dearly wanted to claim the centrality of Both/and. As a matter of structuring these lives, the cultural milieu matters. Relationships, media influence, and the possible discourses that are opened for students open and close possibilities. Why? How does this process work?

5.2   Unsettled by Evolution

When Creationist students were asked to consider a scenario where evolution might be true for them, they shared with me stark and emotionally laden explanations as to why evolution simply could not be true. For this type of student position—"no way evolution can ever possibly be true"—ideas like evolution can prompt a form of existential anxiety by which students "flee" back to the inauthentic safety of doing what one does, and not questioning the foundation of their religious ontology as effected by evolution. This inauthenticity is made possible by the dominant religious discourse having moved away from, and turning a critical eye toward, Biblical literalism some time ago. But therein lies a sociostructural rub. What if one has never experienced theology as either metaphorical, open to change, or something that could be abandoned?

As we have just seen, people do change their ontological position. The student cases I have just presented illustrate scenarios where worldview is in fact changing, or the sociality of their lives is pushing them toward competing ideologies. To apprehend this kind of student "reason" extends us far past the cognitivist limits that many "conceptual change" theorists have been working within. Although leaders in this genre of work (Sinatra and Pintrich 2003) have begun to make a turn toward recognizing the importance of affect and ideology in people's being, their work still assumes a cognitivist approach, not accounting for the larger units of symbolic understanding against or within a gestalt holism by which people make sense of the whole of their worlds. Additionally, as the stories of Cindy, Renee, and Nolan show, the practical everyday reason by which students consider evolution when their worlds are in flux is simply not value free, or easily boiled down to a cognitive "choice."

The social milieu and the intelligibility that the world of social practices lends students were at times seen as being threatened by evolution. Interestingly, in the cases of Cindy, Renee, and Nolan, evolution was both a charged point of discussion for them and a token in larger sociopolitical struggles which are demonstratively connected through their social networks and life worlds, all the way down into their most intimate relations. For a social scientist trained in the anthropological tradition, the deep importance of this shared intelligibility was quite honestly the crux of the matter. Whereas Sinatra Southerland, McConaughy and Demastes (2003) have an inkling of the stakes of this, they run up against seemingly conceptual walls put up by the methodological limits of a commitment to cognitivism. "If conceptual change requires that students compare rival explanations, we argue that such comparisons require relatively open minded, nonabsolutist cognitive disposition when the construct is a controversial one" (p. 525). Our story so far has illustrated examples of how absolutist disposition is created. In what ways are these fostered or changed? Why do religious exclusivists not just slide in an inclusivist outlook? For these types of discussions, one needs to turn toward theorizing the particularities of culture.

To move past the limits that cognitivism prescribes for our thinking, I turn to the work of Swidler (1986) as her work helps illustrate the cultural rationale of students. First, let us think about how cognitivist approaches to some educational problems do not quite articulate the nature of what seems to be going on. Swidler moves past Weber's (1946) classic railroad "switchmen" metaphor of culture where culture is seen as simply switching tracks between the ends of interests. First, Swidler does away with the notion that we are ever "pure" choosers of our actions free of mediation:

> The view that action is governed by "interests" is inadequate in the same way as the view that action is governed by non-rational values. Both models have a common explanatory logic, differing only in assuming different ends of action: either individualistic, arbitrary

"tastes" or consensual, cultural "values." Both views are flawed by an excessive emphasis on the "unit act," the notion that people choose their actions one at a time according to their interests or values. But people do not, indeed cannot, build up a sequence of actions piece by piece, striving with each act to maximize a given outcome (p. 276).

Culture in this way shapes the structure of possible action prior to any choice. Swidler draws heavily here on the concepts of Bourdieu (1984, 1977) and his conceptions of culture as capital, and the theory of practices in which he worked. Swidler's (1986) strategies of action are formed and shaped by the repertoires of cultural action into which we are habituated. "Action is necessarily integrated into larger assemblages, called here 'strategies of action.' Culture has an independent causal role because it shapes the capacities from which such strategies of action are constructed" (p. 276).

With this, we might, in quite the Heideggerian way in the same vein as Bourdieu, see our actions as existing "toward which" or "'for the sake of which." In that, our actions are structured by the cultural capacities within which our lives, our "habitus" (Bourdieu 1977), have been shaped. As Dreyfus (1991) best explains about this view, a large part of cultural dealings *in-the-world* are "withdrawn," in that we are not "choosing" anything, that we are in-the-world in the flow of practices, using the cultural "equipment" or our worlds effortlessly. Although the narratives we spin about the ultimate ends we have in life do not reflect this in everyday life, (as this would ask us to strongly concede the unwilled socialization we have taken part in) the strategies we employ are structured by the social training and capacities of culture in which we reside.

In the case of Creationists, rejection of evolution serves a purpose to bolster or reinforce the systems of cultural meaning making within a Creationist worldview. It is not that a person is prescribed from seeing evolution affirmatively, rather, within the cultured capacities of their life, what function would evolution serve in the cultural equipment and ontology of the Creationist? Creationist rejection of evolution then seems more reasonable when we account for the world in which their practices have intelligibility:

> Strategies of action incorporate, and thus depend on, habits, moods, sensibilities, and views of the world (Geertz 1973). People do not build lines of action from scratch, choosing actions one at a time as efficient means to given ends. Instead, they construct chains of action beginning with at least some prefabricated links. Culture influences action through the shape and organization of those links, not by determining the ends to which they are put (Swidler 1986 p. 277).

So what happens to people who see religion as all or nothing? What is the shape and orientation of a Creationist's conception of all possible theology? Are any other positions *ever* tenable? It might seem pedantic to state in print, but the organizing principle by which fundamentalist religious movements compose their ideology excludes the possibility of other voices or lines of thought:

> Culture provides the tools with which persons construct lines of action, then styles or strategies of action will be more persistent than the ends people seek to attain. Indeed, people will come to value ends for which their cultural equipment is well suited (Swidler 1986, p. 277).

One cannot easily and simultaneously *be a Creationist* and *be prepared to accept* evolution. Taking part in one set of practices excludes others—in this obvious case when firm ideological ends are in question. We only have one bodily *being-in-the-world*. Careers, religious heritage, and relationships are all limiting forces within our lives. How else could it be? This is not to say that the multifaceted, dramaturgical perspective of Goffman (1959) does not apply for Creationists. One can be a mother, a Creationist, and a doctor...but it is weirdly inappropriate to imagine a multifaceted combination of social identities including: mother, Creationist, atheist, doctor, and astronaut. There are limits to the set of masterful practices that we can claim to have, and likely few for which we can claim authentic phronesis. Some social identities, to take them on authentically, require an investment in training and acculturation which by their simple commitment of time exclude the possibility of mastering many others.

Unto themselves, these commitments and the social practices that disclose them "light up" a "clearing" of intelligibility between people, but as the metaphor illustrates, the world we make and exist within has limits. We have a finite time, capacity, and horizon to individually inhabit a cultural world. We have no choice in being born into a world with cultural equipment—a heritage of traditions embodied in situated understandings. We can look at this as our "thrownness" (Heidegger 1962 [1927]), in that we have a cultural repertoire and style into which we are born, which we can build upon, but ultimately cannot fully get behind. As our life opens up a world of new possibilities, it also forecloses the ease into which we might explore others.

In my own case, the abandonment of religion was not met with emphatic efforts to explore other liberally theistic communities or traditions of religious practice. In fact for both science and religion, in my early years, the bar was set so low, and I was socially isolated enough that my childhood interest in science won outright. Moving toward the implications of this line of thinking for education: "One can hardly pursue success in a world where the accepted skills, style, and informal know-how are unfamiliar. One does better to look for a line of action for which one already has the cultural equipment" (Swidler 1986, p. 275). Such lines of reasoning chime with psychological perspectives on divergent thinking that see children imagining fewer categorical possibilities over time through their education (Runco 1991; Ainsworth-Land and Jarman 2000).

How do we build an appropriate toolkit for dealing with evolution education? As Swidler continues, the problematics of Creationist ideology toward both science and religion appears to be especially vexing:

> To adopt a line of conduct, one needs an image of the kind of world in which one is trying to act, a sense that one can read reasonably accurately (through one's own feelings and through the responses of others) how one is doing, and a capacity to choose among alternative lines of action (Swidler 1986, p. 275).

Creationists *can* imagine a world in which evolution works, in that they picture one in which the Devil is at work bringing about lifestyles in which various affronts to their ideology run rampant—homosexuality, abortion, and socialism. When the entire corpus of your cultural world has been shaping you to avoid the "evils of evolution," what possible good can you imagine, and what possible world of practice can you push forward into where evolution makes sense and recedes into the commonplace toolkit of equipment? "Action is not determined by one's values. Rather action and values are organized to take advantage of cultural competences" (p. 275). Quite simply, avoiding evolution, for some, is to commit to sticking with that which "makes sense." As Swidler makes clear:

> People do not readily take advantage of new structural opportunities which would require them to abandon established ways of life. This is not because they cling to cultural values, but because they are reluctant to abandon familiar strategies of action for which they have the cultural equipment (p. 281).

As a reasonable comparative aside, natural scientists who have never been called to account for how they make relative sense of culture and social power within their own lives also show signs of indignity and affront at the possibility that they might not always (*or ever?*) have been "purely rational" creatures. Kuhn's (1970) strongest impact on intellectual thought wrought this kind of reaction from those least likely to consider the contingency of their cultural lives. To acknowledge knowledge as shifting paradigm was to lose hold of scientific *Truth*.

Turning to the experiences of Cindy and Renee, when life was unsettled by major life upheaval, the cultured structure to their habitus opened limited possibilities for restructuring their possible strategies of action. In Swidler's (2001) study of middle-aged, white, Californian residents experiencing upheaval in their lives, this was when new strategies of action could be formed. In Cindy's case, when the course of life was challenged by the stress of an abortion and the reprehensible actions of her community, the discursive spaces in which she lived and shared her life became a limiting factor to the possible strategies of action regarding faith. As she explained, her schedule, her current friends, and her intimate relationships had a strong effect on the direction her relationship to faith was taking. Characteristically, as Cindy's understanding of religion in the USA had been limited by the absolutist liturgy of evangelical Christianity, as her faith faded and began to change, she stumbled over how to even embody her current faith identity: "thinking that there's a higher power, but not knowing what it is." For Cindy, nothing in her cultural capacity and social network of possible discourse was directly setting up a more moderate theological position as an epistemological alternative. But she was not rejecting evolution—she was, due to the contingency of her recent life's alienation from faith, now finding evolution a reasonable possibility.

In almost inverse relation, as Renee experienced the stress of marital infidelity she turned to the church as a form of support and meaning. Having had parts of her prior "world" overturned, her until-then secular identity and understanding of science starts to become reframed by Creationist argument. Critically, the content and message of the faith practice, a church staffed by Young Earth Creationists, become critical to the spin being put on evolution. Theistic evolution was nowhere to be seen.

As Swidler (1986) sees, and Cindy and Renee's lives speak of, the possibility for opening new strategies of cultural practice is stimulated during "unsettled" periods of cultural life. Major life upheaval; birth, death, divorce, among others, offer windows by which new strategies of action are formed:

> In such periods, ideologies-explicit, articulated, highly organized meaning systems (both political and religious)-establish new styles or strategies of action. When people are learning new ways of organizing individual and collective action, practicing unfamiliar habits until they become familiar, then doctrine, symbol, and ritual directly shape action...People developing new strategies of action depend on cultural models to learn styles of self, relationship, cooperation, authority, and so forth. Commitment to such an ideology, originating perhaps in conversion, is more conscious than is the embeddedness of individuals in settled cultures, representing a break with some alternative way of life (p. 279).

When religious ideology is set up as all-or-nothing and the course of life up-tips or throws into question a world of social practices, is it surprising then that Cindy does not know what she believes—that in the practice of her "new" life there is a "nothing" of religious practice? Who pray tell would have structured a Both/and theology, or any other option, for her? Renee, having lived the inverse, and shaped by the cultural capacity of her father's happenstance church affiliation, is now getting "all." Christian theology for her was being structured as an absolute. The social networks and life-world of such individuals structure the possibilities of possible routes that a person might reasonably take. For example, when a person questions or adjusts their commitment to and relationship within an absolutist faith, what other reasonable possibilities pop to mind? The Fundamentalist does not suddenly go "oh, I get this evolution stuff now...now I'll go do this more theologically moderate religious practice instead." There are affective, worldly, and bodily commitments to cultural practices, relationships, and ways of understanding that are not easily just "let go."

Nolan's story provides us with yet another spin on this world of "settled and unsettled" practices and cultural meanings toward evolution. For Nolan, who clearly articulated an increasing detachment from some parts of his faith practice, still turns to the church for moral guidance. "That's the most important thing to me right now, so that's what I kind of go for, and when I do go to church, I listen, pay attention, try and get the morals and values type stuff from it, and not so much try to look to find something to disprove one thing or another." Nolan very much embodies someone Gould (1999) might have hoped to be speaking to—in setting religion and science apart. But as Nolan shared, his narrative was not that simple or in line with scientifically literate opinions toward Creationism. Wishing to "go along to get along," Nolan contends with a slightly different set of culturing influences on his discourse toward evolution—a set of parents diametrically opposed in their views toward evolution, and a set of parents who never speak about science in a way where evolution would ever come up. Additionally, a potential unsettling influence in the form of his girlfriend and her blunt position toward evolution punctuate his life with almost certainly more than the normal discussion about evolution. But as Nolan comes structurally close in his practices to setting religion and science apart, the fact that he willfully takes the Creation Museum's message as a possible "alternative" to consider, certainly will make most scientists' or science policy makers' skin crawl.

What then of the more moderate or accommodating position between the cultural extremes of Creationism and scientism that I saw generally of the Both/and position? Can one count on these folks to be civil toward placating Creationists in civic discourse, or trying, à la scientism, to make them go away permanently? I will need to assess this middle position for its hegemonic normativity, as Creationists likely would disagree with my marginalizing them to the boundary of more "normal" religious practice. This is easy enough to do for a secularist; I do not have a "dog in the fight" as to how politically and theologically liberal or conservative God should be. Any such analysis calls into question and "makes strange" any claim by liberal theists that they have some kind of special access to a more reasoned or truthful position. Returning to Swidler (1986):

> In settled cultural periods…culture and social structure are simultaneously too fused and too disconnected for easy analysis. On the one hand, people in settled periods can live with great discontinuity between talk and action…That is because ideology has both diversified, by being adapted to varied life circumstances, and gone underground, so pervading ordinary experience as to blend imperceptibly into commonsense assumptions about what is true (p. 281).

The liberal theist, our Both/and students from Chap. 4, when asked certainly would not consider their religious heritage and practices to be an "ideology"—for them, that is what "the Fundamentalists" have. Lulled into an illusion that their position is somehow normal and inevitable, for many in this position science and religion are "all good," as long as the extreme is moderated. "Whiteness" in America seems to work the same way…that white people are not "ethnic"… that they could not possibly have practices which when "othered" stand up as a salient position for critique. The normativity of religious practice can be seen this same way. Believers of almost any stripe certainly get their back up when questioned as to the basis of their commitment to belief, because in the cultural practice of faith commitment seems so inevitable, inextricable, and capital T *Truth*ful.

Combining both Swidler and Wuthnow's sociology, for ideology to come to the fore when "unsettled" for the Both/and, is a shade different than for those of the religious exclusivist position. If we believe the rhetoric of Creationists, secular humanism has wrought such an effect on the type of good society exclusivists are trying to build, that one might fairly assess the whole movement as in a constant state of "unsettlement." Agreeing with Swidler and Wuthnow's view, Christian exclusivists are always highly ideologically charged, ever mindful and respective of God's smiting hand. Atheists and agnostics represent another side to this same analysis. A very good Garfinkelian (1960) case to demonstrate why the scientism of so many natural scientists irritates believers in general might be to ask Christian inclusivists: what new practices will they take up when the Christian God is finally removed from its current privileged status and placed on the dusty shelf of the historical pantheon? Talk like that is batted down immediately for the difficult and alienating epistemological lifting it requires. In this way, the liberal theist position advocated by Gould for education is for those so inclined, so naturally "the thing to do" or "what one does," that voices like Dawkins (2006) are then taken to be shrill and downright rude, as they call into question that which must not be called into question. Continuing as Swidler (1986) would see Christian Both/and normativity:

> Settled cultures are thus more encompassing than are ideologies, in that they are not in open competition with alternative models for organizing experience. Instead, they have the undisputed authority of habit, normality, and common sense. Such culture does not impose a single, unified pattern on action, in the sense of imposing norms, styles, values, or ends on individual actors. Rather, settled cultures constrain action by providing a limited set of resources out of which individuals and groups construct strategies of action (p. 281).

Although the life course for the individual liberal theist may slide into agnosticism, atheism, or perhaps rarely fundamentalism, the question as to the cultural legitimacy of the entire theological project as intrinsically important to its constitution will never come into question. Creationists seem "crazy" or "scary" to liberal theists, but structurally they are simply directly competing in the social milieu toward differing ideological ends. As a direct response to the "worldliness" Creationism sees emanating from evolutionary theory, its agenda is to refocus as many of the public as possible on the (as they see it) questionable move Descartes makes during the enlightenment. Science moved scripture aside from its position of authority for all foundational matters and replaced it with *res cogitans*.

There are social costs to abandoning religious faith and systems of meaning. This effect can be seen by more than Creationists imagining the dissolution of their world if we imagine a scenario where liberal theism in America would have to take the prospect of a "godless world" seriously. How would we re-tool the strategies we employ in culture? Are we adequately set up to do this? Are we, à la Nietzsche, living in a period of nihilism? If Christian fundamentalists were to suddenly disappear, how would the civic discourse regarding religion and science continue? Would tensions abate, or would new shades of difference come to typify new struggles? We will explore these themes further in the next chapter where we back our focus out from student experiences and begin to consider the campus cultural environment in general.

Chapter 6
Evolution, the University, and the Social Construction of Conflict

And out of the distance there arose a yell
Ha, ha, said the devil, we're nearing hell!
Then oh, how the passengers all shrieked with pain
And begged the Devil to stop the train

But he capered about and danced for glee
And laughed and joked at their misery
My faithful friends, you have done the work
And the Devil never can a payday shirk

You've bullied the weak, you've robbed the poor
The starving brother you've turned from the door
You've laid up gold where the canker rust
And have given free vent to your beastly lust

You've justice scorned, and corruption sown
And trampled the laws of nature down
You have drunk, rioted, cheated, plundered, and lied
And mocked at God in your hell-born pride.

Hell Bound Train (excerpts)
American traditional

6.1 The University at Large

I have examined students experiencing existential anxiety in the face of evolution, the ontological positions which typify all students regarding evolution, and have begun to examine cultural structures that variably prepare students to converse regarding evolution. For some fundamentalists, worldview *can* change in ways that evolution becomes tenable, although the process is neither value neutral nor lacking social costs. I now back up a step and consider the campus cultural environment in which conversations regarding evolution play out—explicitly, implicitly, and not at all.

D.E. Long, *Evolution and Religion in American Education: An Ethnography*,
Cultural Studies of Science Education 4, DOI 10.1007/978-94-007-1808-1_6,
© Springer Science+Business Media B.V. 2011

Fig. 6.1 A "Cemetery of the Innocents" makes its spring seasonal appearance on Mason-Dixon's Campus, protesting legal abortion

Fig. 6.2 Its ideological counterpart elsewhere on campus. Sign reads "Women's rights grow with choice"

I made my entrance to Mason-Dixon State as the university was preparing to hold a Mock Trial concerning evolution and Creationism. Although a fortuitous event to dramatize the starting point of my project, it also crystallized what boils beneath the civility of many college campuses. Ideologies are embodied in people who then project worldviews out into their respective fields of social significance. Faculty, who by the dominantly liberal political discourse of the academy, often do not have to take religious or politically conservative views seriously as they go about their own research. For natural scientists especially, as I have commented elsewhere regarding Bloom's view on this, politics is a necessary evil that must be addressed when the funding lifeblood of science is threatened (Long 2010a). But to do this efficaciously, many scientists are likely hampered by a poor vocabulary for how people talk about/with faith as Ecklund (2010) has shown.

Regardless of how scientists see the politics of this process, many students find no difficulty in associating evolution as a turning point for political ideology, which in recent years seems to be increasingly distilled by apparently incommensurable philosophical foundations. In its mundane form, student imagination about the purpose of higher education extended the utilitarian K-12 schooling metaphor of education for four more years, until one could get a job and enter "the real world." A few earnestly described college as feeding a life of the mind. Yet another few, all religious exclusivists, saw the university as seat of one of the greatest "evils"— secular humanism.

Was this Mock Trial an exception on an otherwise politically tepid campus? As my preliminary investigations with campus administrators and student organizational leaders explained, such events were not exceptions. During my time at Mason-Dixon, many other issues that alienate religious exclusivists and emboldened more inclusively minded individuals were played out in the public spaces and discourses of the campus. Tangibly, a yearly antiabortion display appeared, countered by one supportive of women's choice (Figs. 6.1 and 6.2). This latter counter display was a more recent addition as in the very recent past the first iteration of the antiabortion display was destroyed by a Mason-Dixon literature faculty member and some of her graduate students.

6.2   Evolution and Creationism in Public Pulpit and Print

Mason-Dixon, like many American college campuses, at least in Middle America, is a natural attraction for all manner of "enthusiastic" public speech, often from externally based advocacy groups. Frequently, speakers come from religious exclusivist groups. During my time on campus and right before, some exceptionally vivid examples appeared. Evangelical "campus preachers" were a regular attraction on the quad at Mason-Dixon: John McGlone and Kerrigan Shelly of a Pinpoint Evangelism and a "Brother Rick." Mason-Dixon's visits were typified by students being warned "all sinners are going to hell" as they walked by, and often being singled out as they did so. Evolution almost always was a key point of contention for them in their public rhetoric.

The campus newspaper was a lively space where such "battles" took place. Prompted by the actions of the biology department posting a disclaimer to the Mock Trial on their department web site, a student writes a pointed editorial to Mason-Dixon's student newspaper. Excerpts that delineate the crux of his argument include:

> The way I see it, the problem is twofold. Even in a religion whose essence demands faith without proof, mankind has some innate desire to prove that God exists to settle that empty feeling we get in the depths of our stomachs sometimes called doubt. And secondly, when things start to go wrong (whether it's our economy falling down the stairs, or just a few too many rainy days) people tend to want to blame something. I'm glad this time that it's science, and not a person or race, but that doesn't make it right.
>
> …It's a problem for those in power who are both scared for their country and their children, and who think that forcing religion into the classroom is the answer. I'm a firm believer in science explaining how and religion explaining why, and I wish more people would do the same. They can believe what they want, but leave it out of the classroom.

Naturally, given all we have covered regarding science, religion, and the civic discourse regarding evolution, an answer was supplied. The point of origin was a bit of a surprise though. A tenured faculty member of the Mason-Dixon psychology department writes in excerpts:

> First, I agree that faith and science need not be at odds. Indeed, many science pioneers were religious. However, the creation / evolution issue is often presented in a biased way. It is described as Creationism vs. evolution as though the creation side is more ideological. Or it is described as science vs. creationism as though creationists are against science. Creationists are not against science. They simply think that the theory of evolution is not good science…Science should follow the evidence wherever that leads and if it leads to God, that does not make it unscientific.
>
> …Second, it seems that some people try to leave the impression that Creationists are few and that they are ignorant. But anyone familiar with the polls on this knows that nearly half the people in the U.S. are Creationists, including many who are intelligent and knowledgeable.
>
> …Third…I have noticed that many people seem to have the idea that faith means believing something for which there is no evidence. Perhaps in some religions that is correct, but biblical faith refers to trust or belief based on testimony or other evidence.

Although the student editorial that prompted this reply was not of an unusual tone for a college campus, the tone of the reply was. Was this faculty member writing as Devil's advocate, or in fact did he believe himself to be alienated as he was a Creationist? There were certainly a group of students who might agree with his viewpoint; however, what might be the view of the faculty?

6.2.1 Creationist Faculty: Come Out, Come Out, Wherever You Are!

That a public college faculty member might be a Creationist comes as a surprise to many. Given that there is really no good reason why the same symptoms of poor science literacy or fundamentalist religious commitment would not extend to university

Fig. 6.3 Flyer posted at the University of Kentucky

faculty, it should not. Unless you are in the natural sciences, graduate work does not in most cases overlap with these domains. As Kurt Wise most famously showed, one can gain the highest credentials in the natural sciences from an oracle of evolutionary biology (Gould) all the while being a Creationist (Chapman 2001). That is not to say that I expect there is a high percentage out there. But in my own graduate work at the University of Kentucky, one by one a handful of Creationist faculty popped onto my social radar, one quite interestingly working within the biology department itself.

While writing this book, I paused to attend a public lecture advertised for my own campus which naturally piqued my interest (Fig. 6.3). Given that the university was about to have a symposium in honor of Darwin and the 150th anniversary of the *Origin of Species*, I presumed that guest speaker David Eakin would strike an ecumenical tone between science and religion for the Bible Study Club. As I sat down within the lecture hall, UK biology faculty member Jim Krupa joined my side. Through a torturously long exposition of his obtuse interpretations of classic Creationist arguments, it became clear that Eakin was not there to adjudicate Intelligent Design as a social phenomenon unworthy of science. He was there to critique it in favor of Young Earth Creationism! The most worrying part was that Eakin was the coordinator and primary instructor of his university's nonmajor's biology courses. Dr. Eakin teaches at Eastern Kentucky University, where each semester he delivers biology course content to hundreds of students, with cadres of future science teachers among them.

Back at Mason-Dixon, I arranged to meet with our letter to the editor author Dr. Yates, a psychologist specializing in social psychology and the psychology of religion. I confirmed his interests in Creationism. Like many Americans, he is unconvinced that the evidence supporting evolutionary theory is adequate. Like each person I have ever encountered who shares this particular strand of skepticism toward evolution, he is also a religious exclusivist. As a Church of Christ member, similar to five students within my interview pool, he is in the evangelical business of attempting to bring all Christianity under one roof, and read to it a literalistic Bible. As we speak, his own social circumstances are both interesting and resonant, given the lives of students I examined in Chaps. 4 and 5. Growing up in southern California a secular Jew, he eventually "finds Christ" while in graduate school and joins this socially marginalized form of religious practice. Of note, this is the faith practice of his wife that he meets during this process.

As in all my interviews, as one identifies themselves as a Creationist, I first have them delineate the inadequacies they see in evolutionary theory. After this, I ask them to imagine what it would mean for them to find that adequate evidence suddenly *was* there and evolution had in fact happened. What then would this mean for them? In Dr. Yates case, as in many other interviews with Creationists, the phenomenology of such lines of questioning makes for interesting exchanges. Flushed and rapidly turning his wedding ring, Dr. Yates replied:

> That's a great question. There's no doubt that that would be problematic for me. If that sort of thing were to occur…that would force me to rethink how I look at the scriptures…I don't think it would cause me to abandon my faith, because as you know there are all sort of people out there who have all different kinds of ways of looking at things…I would have to give a lot of thought as how to reconcile, but as for now I think my worldview and all this stuff is very consistent.

As with all my interviews with Creationists, the discussions at times graded toward metaphysical boundaries, an area day-to-day discussion really never goes. Dr. Yates saw this line of questioning for what it was. Given the requirements of university work, one must regularly negotiate multiple religious perspectives. Nevertheless, in his exposition of the inadequacies of evolutionary theory, he was by his own admission a novice regarding the geological context and content of the fossil record, citing Darwin as seeing the fossil record as a liability. As I wrapped up my interview with Yates, he made clear that in his written advocacy work for Creationism on campus he was not alone. In times past, Yates and two other faculty members of other departments had written supporting Creationism some 10 years before. Upon investigating this, one faculty member was a music professor and another was in business, having since left for another university.

My experience with the Mason-Dixon psychology department also directed me across the hall toward a Dr. Bollea. A bit ironically, along with key individuals in the biology and philosophy departments, Dr. Bollea had just set up an evolutionary studies minor on the Mason-Dixon campus. Himself a matter-of-fact atheist, Bollea saw the evolutionary studies program as a timely means of pushback against what he felt to be a growing antievolution trend in students influenced heavily by their local churches. Across campus, I met with a Dr. Howard who taught a history and philosophy of Darwin class as part of this minor degree program. Also an atheist, Dr. Howard was interesting in that he had in almost inverse relation to Dr. Yates by religious practice growing up in the Anabaptist tradition in Southeastern Pennsylvania, actually quite close to my childhood home. During our brief interview, he also explained that his department also has at least two faculty members who are Creationists, both teaching classical and ancient philosophy.

6.3 Evolution in the College Classroom

Shifting from the phenomenology of student-centered perceptions of evolution education, we recenter on the social practices of the classroom. It is within this setting that the often nonverbal contention that I have described plays out. During the course of my time at Mason-Dixon State, I followed three introductory biology classes (one major's course and two nonmajor's) where evolution would be taught. Throughout the course, I would take notes on the classroom practices with special attention paid to the types of discourse used, and with particular attention paid to which topic prompted dialogue, and which topics met with silence from students. My attendance in class was briefly announced at the beginning of the semester, which garnered a glance or two, and thereafter no response. Sitting in a rear corner of the room, my time-series notes recorded the outline of the lecture (the classes were almost entirely teacher-centered lectures), the type and frequency of student interaction, namely, questions, and which class topics prompted them.

Generally, the classes were free of discussion. When students did pose questions, they centered around two issues: the explication of complex topics (the chemistry of photosynthesis), or quick exchanges regarding controversial topics. In total, two sections of the classes I watched produced five student questions each over the ten class periods; the remaining class, taught by an enthusiastic Dr. Green, produced 33 questions. It was also within this more interactive class that the contentious nature of evolution, or other topics such as genetic engineering, prompted social tension. In the most contentious scene within this class, a student strongly objected to any prospect of genetically engineering plants or animals as "playing God." Genetic engineering and cloning together garnered 19 questions, each within a burst of about 10 min. These two spurts represented almost half the questions asked or prompted during approximately 30 h of class contact time. Active student participation brought contention to the surface, where student silence and passivity in lecture did not.

At the onset of the project, I had hoped, or expected, that lectures that covered evolution would prompt lively discussion. In a surprising move, in the same classroom that produced the largest amount of questions in total and the buzz around genetic engineering, discussion of evolution was carefully shaped to tamp down potential discussion. Prompted by a distinction Professor Green makes, a student asked "so why is there a difference between micro and macro evolution?" It is important to note that Creationist arguments often pose unorthodox conceptual splits not recognized by most scientists. The rhetoric of Creationists is to stress that "microevolution" but not "macroevolution" has taken place, a distinction that orthodox science has recently refrained from emphasizing due its propensity for generating potentially confusing and politically charged environments. As Dr. Green Responded to the student: "They like that because some people do not like the definition of macro-evolution," she hesitated and redirected into a lengthy and parenthetical justification:

> Some people—because we can see micro evolution in terms of bacteria…we can watch bacteria evolve…in a number of hours. We can see their allele frequencies change. Now… macroevolution is a little more complicated and it takes a little more time to observe—and for some people it goes directly against their religious beliefs.

Dr. Green next completed the argument. "Macro evolution is seeing evolutionary change at the species level or higher. So what we're talking about…the accumulation of these allele frequencies…multiple phenotypic changes may result in a new species."

Given the slightly oppositional framing of the premise because "it goes directly against their religious beliefs," I was not surprised to find a silent room staring back. Rather than engage in a dialectical discussion of the basis of this rejection by "some people," the issue is set aside from discussion, letting the likely large minority of students who hold such positions to quietly stew. We might not fault this faculty move, as like most everyone else I interviewed, and I will discuss this further, their own intracultural religious literacy was not the greatest.

In an attempt to then integrate micro- and macroevolutionary processes, the professor showed a video on sickle cell anemia. In showing this kind of change in allele frequencies in humans, the professor hoped to demonstrate such microevolutionary change in humans. In all, other than this careful control of classroom rhetoric, discussion of evolution was mainly marked by student silence, something faculty interviews would confirm as not out of the norm.

6.4 Course Faculty and Department Practices

Although my experience conducting this research was by and large a pleasant if at times eye-opening one, some faculty were cagey at my presence to say the least. Perhaps expecting that I was some sort of secret-agent Creationist, Dr. Russell had a distinct tone of hesitation at the prospect of an interview. Part of my research design which I would carry over from students to faculty lives would be to have them reconstruct their own histories with science and religion, and the tension, if any, between them. As part of this, I wanted to get their sense of when they came to understand evolution as a controversial concept in American society. The tension that Creationists have wrought is palpable as Dr. Russell became defensive at even the framing of my question. As is clear in what follows, there is a good risk of speaking past one another when one engages in such conversations. Immediately, my interest in the social controversy surrounding evolution prompted tension:

> I…*hesitating and thinking*)…actually don't know that it *is* controversial. And I don't know I've ever thought it to be controversial. I think a lot of people try to put that label on it….but there is no controversy—as far as I can see.

Sensing this was going to take a bit of work, I asked him to imagine a Martian dropping down examining social controversy…"but" (*Russell interrupting*) in the USA…"but there *isn't* a problem with evolution" (*Russell emphatic*). Explaining that I am looking at disagreements regarding evolution sociologically, I asked if Russell is uncomfortable with me framing evolution as a social controversy:

> I think that there are beliefs...right?—and then there's science. Science is not a belief system and the people that have a belief system feel that they think this is the direction that the earth went. But that's strictly based on a literal belief in the Christian Bible.

After discussing that abortion in the USA could be seen the same way, as a social controversy, regardless of one's opinion of the practice, Russell sighed with resignation: "It wasn't really an issue in my home state."

Illustrating how biologists themselves might be a poor read of the extent of Creationist sentiment in society, when I later speak with Dr. Essex, Mason-Dixon's science educationalist in the college of education, he tells a completely contradictory story. Having taught high school in the same "home state" as Dr. Russell (in the upper Mid-west) prior to his move to academe, his science colleagues brought Creationist literature to be shared with their biology classes when the evolution unit was covered. As he described, this in part led to his hasty exit from this position because only one other colleague felt strongly opposed to the presence of these materials as he did. This was but an hour drive outside this state's major urban center in which Dr. Russell grew up where evolution "wasn't really an issue."

A few minutes past the air of tension I sensed Dr. Russell and I producing, his experience with antievolution sentiment began to unfold. Despite the semantics of "controversy," something was certainly afoot in his background. Talking about his experience in helping at his child's school:

> I would bring animals in and talk to the elementary school kids and you know...I'd hear later that I used the 'e' word...and they were told as teachers not to use the 'e' word in the schools...so that was really...I was like, 'you've got to be kidding'.

I asked him to pin down where he sensed this sentiment to be coming from:

> It was something with their school board...it was sort of an unwritten rule coming down from, I don't know—the superintendent...teachers...or the principal? I mean they were trying to change the textbooks....if it wasn't for the separation of church and state, it would have gone through.

During this time, Dr. Russell was an adjunct professor at a university in Middle America while finishing his own graduate work. In this case, his university faculty colleagues took an oppositional stand protesting the move to cut evolution from the curriculum. "The faculty at the local university got together and took out a full page add in the local paper, and said this is our statement, and these are our signatures backing this statement."

Dr. Russell, who now mainly teaches upper-level courses for biology majors, talked about the type of flak he received concerning evolution while teaching as an adjunct during this period:

> I had a student who had a question who emailed me. I emailed her back incorporating evolutionary principles in my answer to her question. But this was to her husband's email account which is what she used to email me—but *he* emailed me back and he indicated to me that 'of course he knew, he had extensively studied evolution—and he knew for a fact that dogs did not evolve from cats' and so...(*laughing*)..I emailed him back, 'You are correct'!

Since arriving at Mason-Dixon he has not felt much active resistance, but in the case that he shared, when it does surface, it is dramatic. "On one occasion I did have a student stand up and say something literally religious and leave the room." Dr. Russell details that he was "giving facts," and explained to the student that he was "sorry you feel that way." As he explains, the student muttered "I can't take this anymore" and walked out of the room.

This type of scene was not unfamiliar to me. During a preliminary study for this book at another university, a more extreme example of this same phenomenon occurred. While this professor directly addressed the historical evangelical and Southern Baptist resistance to evolution in class as a teaching tool, nearly one tenth of the lecture section quietly got up and left class. Of three sections observed in this early research, in two sections, at least 10% of the class left during the explicit discussion of the sociohistorical roots of antievolutionary sentiment—25 of 250, and 43 of 400 students in each of two sections. As this professor discussed with me, this was not unusual, but he had never had means to quantify this. As my interviews with this faculty member would detail, over the years a small minority of students would openly challenge evolutionary science—with the extreme being students who would "stand up in the lecture hall and yell at me that I'm going to hell."

Perhaps being on a plane of wisdom acquired with many years of experience, Dr. Wayne struck a much more dialectically engaging tone toward Creationists who popped up in his classes when they spoke to him in confidence. Himself a Christian as he made clear during his earlier Pascal's wager discussion in Nolan's story, Dr. Wayne had some deeper experiential insight regarding the effect of evolution on Christian exclusivists. "I had a student who was in a fundamentalist group and he was getting pressure from his family. It was not exactly being shunned…but that's what it was approaching." Dr. Wayne and I had immediately struck a rapport due to similar intellectual trajectories, life successes and particular failures, and had both grown up in rural, agrarian areas. As he counseled this particular student about pressure from home: "That was their major concern…how even being a science major affected their standing with their family, their church and their social groups." Dr. Wayne was a useful interpreter between the world of science and the world of religion as he had ensconced himself at Mason-Dixon as both a founding member of the department, and having made community ties in the rural county away from the campus where he lived. With this experience he saw, perhaps like no other, the town and gown issues that made Mason-Dixon a source of resentment to some folks. The tension was clear enough:

> I think a lot of the reason that many of these people won't talk to us is that they think we're all atheistic, we only know science…but many of us are not only members of churches but active members of churches.

This polarization has become especially pronounced in the region with the Creation Museum opening its doors. "They view the Creation Museum as an affirmation of what they've been doing and they view the university as an adversary," Wayne stated. But like some of the students whose worldviews were in flux in Chap. 5, higher education in part fostered possibilities that the institution opened up.

As Dr. Wayne relayed, confirming the social cost of worldview change prompted by education: "That young man ended up going to medical school and going into medicine…but he still doesn't have contact with his family…not because he's in medicine…but because he believes in evolution." Again with wisdom of experience not easily nor quickly taught, "It pointed out to me the risks that [evolution] posed for some students. They risked disenfranchisement…to be any kind of intellectual in a family like that posed risks for their family and any kind of those social groups."

6.4.1 Transmission (Theory) Error

Faculty who taught at Mason-Dixon felt what must be common to any group attempting to do a survey of a field in one class. As Dr. Green expressed when she saw the syllabus for this class upon arriving at Mason-Dixon: "holy bajolies!" Although the course was not unusual to my eyes in its breadth or depth of content, there was a general sentiment that the nonmajors class had been for lack of a more polite term, "dumbed down" to the bare essentials. As Dr. Fleischmann saw it, the nonmajors complaints about "how hard the class is" were an indicator of problems afoot with student achievement generally. As he explained, if these students were to realize the gap between the nonmajors and majors course requirements, "they would have been like—totally destroyed."

Following a form practiced throughout the majority of university science classes in the USA, almost every minute of every class was teacher-centered lecture. It may seem a bit odd to point this out, but it is important to note. Given the dominance of at least a rhetorical commitment to teaching for active and authentic learning in best practices K-12 education, and regardless of the efficacy of either, higher education science is dominated by didactic pedagogy. The laboratory sections, which my observation and interview schema mostly precluded me from participating in, would be the sole respite. This issue was not lost on the department chair, who explained to me that Dr. Green had been brought in specifically to address these kinds of concerns. Her own graduate work focused on more active types of university science teaching and learning. Although my observation schedule may have not seen this fairly, Dr. Green was the only faculty to engage in active learning techniques, for a total of 30 min during my 30 h observing the biology sections.

Didactic teaching works well for the small minority of people inclined toward this type of learning style. Although not a focus of this book, it may well be the case that many in the sciences find this effective due to aptitude or stylistic comfort from having been socialized in this way. I am one of those who can listen and absorb. It is also the case that the ideologies of worldview do not discriminate based on learning style. This transmission theory of education (didactics) has profound limits when we consider what exactly we mean by "education" for students. As a broad program, the AAAS *Project 2061* goals are clear, in that they are attempting to effect scientific literacy through the fostering of "scientific habits of mind." Part of these habits includes an openness of skepticism toward knowledge which is always possibly open to revision. If not already clear, the absolute epistemology of religious exclusivism will not easily allow the undermining of some areas of knowledge. In a very clear way, for both scientific positivists and religious exclusivists, they both seek a *"Truth."* So given the competing ends each might serve (uncovering *Truth* and glorifying *Truth*), is didactic learning effective in these regards?

Dr. Fleischmann had finished a lecture one day, shortly after I had just finished my first set of interviews with all of the students in my interview pool. He was passing back test results from the unit that covered evolution. Beaming, he offered uncharacteristic praise to the class in that they had, to his pleasure and surprise, topped the average of past classes and other sections by more than a letter grade. In fact, he had a number of students who with the addition of a small amount of bonus points achieved a slightly higher than perfect score on the exam. As he handed back exams, he held the very top scorers until the end. Pausing to acknowledge the efforts of each one, they one by one were members of my interview pool—one Creationist, one atheist, and the last and final student, another Creationist who had just days earlier made clear that "we didn't evolve from monkeys." For many science educators, these cases should not be a surprise. Carefully considering this scene, the skills one needs to pass a test and to take a text as literal *Truth* might actually be in the same ballpark.

6.4.2 Have Humans Evolved?—One Would Not Know

Given the focus of my critique on evolution education and where I see it breaking down, one might be tempted to see my overall assessment of Mason-Dixon's work as lacking. That should not be the case. Regardless of a student's religious disposition, there was an overwhelming majority of students that felt their professors had done a good job. I concurred with them, except on one major point. Human evolution was curiously absent from mention. As Dr. Wayne explained when I asked if they mentioned human evolution:

> No…no…what we want to do is introduce the concepts without getting polarized from the outset. They all come in with ideas and thoughts and philosophies and beliefs…and if they see on the syllabus—'human evolution'— they'll say, awe shit… I'm not going to take this course.

Although it could be argued that human evolution might not need to be covered in an intro biology class, such an argument misses a point. For many university students, such intro science classes are the last best chance for evolution education. Prior to their matriculation on campus, human evolution was almost never taught. The other places it might likely gain mention, a paleontology class or an anthropology class, are not classes which the majority of students take.

As Dr. Fleischmann saw it, human evolution was not worth the hassle versus what he saw the department gaining. Fleischmann had been the leading instructor for the nonmajors course for many years. This common course had many sections due to its demand in the general studies curriculum. When I asked him to explain their rationale for not including human evolution:

> (*long pause*) One of the things that…if we had to sum up what we thought students were thinking of [human evolution] was 'monkeys into people'. What we found was, when we taught just natural selection, this was all stuff they'd never heard about before. We really wanted to center on that, rather than sort of …you know, anybody is welcome to go into that if they want to.

As I knew from my preliminary interview with Dr. Fleischmann, and as he had implied, "[for] anybody wanting to" teach human evolution, it had been a sticky issue in the department. As he continued about his own exclusion of human evolution:

> I don't, because I feel it gets people's dander up. And once they put up that wall, I lose my teaching opportunity. So I feel I would rather use my teaching opportunity to initiate these questions in their minds. Here's this evidence and here's how things could work, and here's how things could change over time. I then have no doubt that they're applying this in their own minds to thinking about chimps and humans and things like that. You don't need to do a direct confrontation with them. In the past I have brought up Creationism versus evolution and I found that that was detrimental…that that actually shut them down more than if I don't mention it at all.

As his narrative unfolded, there were clear events in his past which may have instructively shaped his exclusion of human evolution in class:

> I would have students that got huffy and be very confrontational asking questions in class. (*Sarcastically*) They would accept…I could have made up any story I wanted to about genetics—human and pig hybrids and they'd believe me. But bring up human evolution and…(*shaking his head*).

Perhaps not indicative of the primary concerns he has today, Dr. Fleischmann explained that course evaluations would come back with extensive comments. When he used to discuss human evolution or directly discuss Creationism and evolution, evaluation comments would grade toward his being "closed minded." As he explained, he now looks toward these fewer but still present comments as a source of "shits and grins." When I asked him the effect of these direct challenges to his classroom when he did include human evolution, he explains: "It disturbed me. And it made me not want to do it anymore. I could say that (their complaint) was for a philosophy class, not for a science class, but that wasn't acceptable."

Even the most recent hire to the department came to quickly understand the politics of evolution. As Dr. Green recalled, "It came up in my interview. They asked me on the phone interview about my views on evolution. They were trying to 'ask me about it but not ask me about it'." When I asked her if she has since learned why they asked this question, she explained: "I believe there were some people in the past who were not teaching evolution in the intro class…they wanted to make sure that a new hire was on board." I pressed for specifics. "I was told by the chair or maybe Dr. Fleischmann that we have the set curriculum because sometime in the past people maybe hadn't been covering everything they should be covering."

Dr. Green was closing in on the end of her second year as a Mason-Dixon faculty member when I asked her to assess her take on the department's treatment of human evolution in the classroom. "We 'surface' evolution, we do not go in depth there… really, [human evolution] is not addressed. And to some degree I think that's perfectly fine… because really, that is a really recent, itty-bitty part of the whole story of evolution."

Again appealing to the issue of breadth, human evolution simply does not make the cut:

> If you have more time, I think it's a very interesting thing to go into. There's a lot of evidence out there and even in the last 20 years there's just been so much more collected. But for these guys, we don't even get into that with our upper level biology students. It's covered in the book a good amount but we don't have enough time to focus on it.

Perhaps given that many scientists have a hard time conceiving of ways that anyone could actually be a Creationist, a primary issue for Creationists is left off the table of class discourse. For Creationists and religious exclusivists, it is essential to see the primacy and centrality of humankind as the product of a special creation. As organizations such as Answers in Genesis stress, the foundation for the entire project of the enlightenment is at fault in moving the Bible from its privileged position as central and ultimate arbiter of *all* issues. In this light, when Dr. Green excused human evolution's absence and sees microevolutionary mechanics as sufficient to ground students as creatures of the earth, the disconnect is almost painful. "I think it's more important for them to understand alleles and allele frequencies changing over time." As my student interviews with Creationists made clear, allele frequencies could change *ad nauseum* without Creationist students building up the more complex and temporally abstract concept of macroevolution. Although the most effective way to teach about evolution is still in question, and one of the motivating rationales behind this project, omitting human evolution has clearly had its effect on one Mason-Dixon faculty's outlook. "Yeah, I think it's one way to get around having so much uproar in the classroom."

6.4.3 Who Teaches Whom Evolution?

As became clear to me, experience and breadth of knowledge had a lot to do with one's comfort in handling conflict regarding evolution. A bit of extroversion may help too. As Dr. Wayne explained from his years of teaching:

> As one student told me—a long time ago…'my minister told me you're going to try to change my beliefs'…and I said wow…if only I could be so powerful. But as I show them through my teaching—we have the empirical domain and there are beliefs. And the worst we could do with that is have a verbal disagreement.

Contrasting this are Dr. Green's feelings. With just a few years' experience teaching in this contentious territory, she appeared a bit negative: "I've increasingly become more uncomfortable teaching it…because I'm worried they're going to do something…like they're going to ask me questions about it that I can't answer."

As Dr. Wayne explained though, the growth of Creationist attitudes bolstered by the opening of the Creation Museum have, in tandem, sparked a bit of a call to arms:

> I think it's made us realize there's a problem. We tend to get wrapped up in academia unless we're out in the real world doing what they call outreach now. I think a lot of faculty are afraid to get involved…but not from the standpoint of physical harm or tenure and promotion issues.

Dr. Wayne's solution was tempered by seeing the growth of his institution begin to shift focus from undergraduate education toward more research reassignment time. He worried about what this portended for faculty/student relations. As he saw it, the remedy was to have more faculty engaged in teaching a broader repertoire of their discipline to a broader range of students on a regular basis, as demanded by the introductory courses:

> It gets back to having all faculty involved in the intro course. Faculty are afraid to do that because they are going to have to visit sub-disciplines they're not as familiar with…and they'd rather be wrapped up in the labs with their own research.

Unlike Dr. Green, who specialized in science educational pedagogy as part of her training, when I asked Dr. Fleischmann what his teaching the nonmajors had to do with his research interests, his response was "absolutely nothing."

These types of sentiments are part and parcel of a system of academic social capital which increasingly values research over teaching. Like most other social fields, academics are no less susceptible to the general pressures of an increasingly market-based system where the social capital of self-promotion by one's research "program" is the utmost concern. Bieber's (1999) analysis of this for faculty life is no less resonant than 10 years ago. As Dr. Wayne feared, newer faculty (Dr. Green, a thankful exception) increasingly demand lighter teaching loads so as to balance their research agendas. Research certainly has its central place in academe and specifically laboratory and field sciences. But at what general cost?

Given what had already been explained to me regarding hesitation in general, regarding evolution, and avoidance of human evolution altogether, the following scenario did not exactly add up to evolution being cleanly or clearly presented. Faculty had intimated that evolution had been entirely avoided by some part-time/adjunct faculty in the past. As I will discuss in the next chapter, the quality of prior evolution education brought to college by the cross section of students I spoke with was terrible. So what was going on? Again, with more years of experience, I turned to Dr. Wayne's institutional memory: "We've had adjunct faculty here…simply because we needed bodies to teach the courses…" He trails off and then comes back, "and I'd walk by the classes hearing them teach genetics…and I'd go (*wincing*) 'oh God no—I don't want to hear this'." I clarified as to whether there were some who were in fact not teaching evolution which he confirmed. As he explained his understanding of the situation:

> I knew of one or two…they were high school teachers. They didn't know evolution. It's the same reason why they shunned molecular biology…they were uninformed in that area. That's part of it…some of it could be theological…some of it could have been a lack of

understanding. But it's all a lack of understanding—they couldn't explain it in a way that allowed their beliefs to still exist. And so they had personal conflict with this stuff and even though it was part of the curriculum, they were very uncomfortable presenting it in the classroom.

As I was naturally curious as to the back story of this situation, I asked Dr. Wayne how he confirmed this:

With one particular individual I did. He was the science department chair at a local high school, and was related to someone here on the administration. We talked about this. His concern was it would reflect on his position at the high school if it was known that he was advocating the evolution position . What it really came down to was he didn't understand it.

As Dr. Wayne's insight clarified, one of the first places the general studies portion of university science education gets compromised is the nonmajors science experience. In a separate part of his interview, Dr. Wayne spoke affirmatively about the equally compelling need to have the most highly competent people in the majors classes. When I asked him where and how the department has used adjuncts and part-timers, he was definitive. "The non-majors course, that's where the adjuncts teach. We try not to have any major course taught by an adjunct faculty member." But as his own views and departmental practices belied, at worst evolution was avoided, and at its best it has definite room for improvement.

6.5   Evolution Outreach and Pushback

6.5.1   Darwin Day

As my interview with Dr. Russell made abundantly clear, the rhetoric associated with evolution was highly charged and laden with power relations. Given the politics of "controversy" understood well by Dr. Russell, my attempt to move the analytical frame toward social conflict met with strict Foucaultian disciplinary measures. "There's no controversy!" It was as though by my not invoking conflict that Dr. Russell symbolically kept Creationists at bay. But Dr. Russell and Mason-Dixon's biology department actions belied a deeper reality that they might find hard to swallow. Their fight against Creationism was at best a stalemate. Without further advocacy, the department would be handing the next offensive move to the political milieu of activist evangelical church congregations in league with the strong influence of cultural actors like the Creation Museum in the region. Across campus from the biology department, Matt Leslie and the Campus Christian Center were actively producing antievolution programming. The campus sponsored a Mock Trial for evolution in which the biology department was only cursorily acknowledged. Even the politics of what constituted a museum was a rhetorical chip. As Dr. Russell insisted in our interview, the Creation Museum was not a museum; it was the Creation "Center" as he put it.

When the college held its Mock Trial, the department at least protected its borders by cordoning off the rhetorical territory of biology, if the college would not. Additionally, Mason-Dixon and Dr. Russell were in their first few years of taking activist steps to at least attempt to change the rhetoric of the local schools. In this spirit, Darwin Day was born. As Dr. Russell saw it, Darwin Day would move support for the curricular core of evolution from public schools to the college campus where the rhetorical center toward evolution was squarely supportive. A typical Darwin day would take a whole grade of high school or middle school students and divide them into differing focal areas on evolution: evolution of microbes, invertebrates, vertebrates, homologies, sympatry and allopatry, etc. As this grew, a local, privately run naturalism center known for their public outreach joined in, and the program grew to a two-site day of education on evolution. Darwin Day has continued to grow and was in its fourth year during my time at Mason-Dixon. Of course, behind the sunny picture of Darwin Day hides the ever-present social pressure and resistance of Creationism. Although there was much interest in schools participating, some clear indicators of the politics of evolution in school administrations cropped up. In an egregious example of one district's response, three separate Mason-Dixon faculty shared with me the following message with indignant relish:

Dr. Russell,

Thank you for the offer to attend Darwin Day. After talking with school personnel and district personnel, we have to decline the offer. Our review of the district curriculum and our standardized test scores do not indicate a need for devoting the amount of time our attendance would take from student instruction in other classes. Second, the topic appears to be controversial and not appropriate for our conservative community norms. Thank you for thinking of us. Have a great week!

Cindy Dolty
Director of Curriculum, Belton County Schools

Rather than dance around the issue of antievolutionary attitudes being perceived as part of a political ideology, Ms. Dolty comes right out and confirms it. Strangely, the point that evolution is part of the Belton County's state curriculum is missed. Or perhaps the fact that their community has "conservative community norms" overrides this as a matter of "appropriateness."

In the early stages of my developing Mason-Dixon as a field site, Dr. Green was forthcoming about some other wackiness going on in local schools prompted by actions which stemmed from Darwin Day. As Dr. Green understood, the Answers in Genesis Ministries, who run the Creation Museum, was granted brief access to a public school's science classes as part of an "equal time" argument—apparently resulting from the school's participation in Darwin Day. Taken aback, I double-checked what Dr. Green had just illustrated for me. Equal time? *MacLean vs. Arkansas* had settled this issue in 1982 (Overton 1982). Now, approaching 30 years on, schools still violated this federal ruling. I naturally wanted to get the backstory to this.

Nolte County was just a short drive from Mason-Dixon. Although Mason-Dixon straddled an urban–rural boundary, Nolte was rural—no large cities, low population density, and dominated by agrarian lifestyles and Christian exclusivist church congregations. As Dr. Russell explained, the trouble probably began with the first official notice of the trip. "Word got out because we advertised it as Darwin Day. Parents had to sign off on the permission slip." A science teacher at Nolte County Schools had arranged for the trip, and the event went off more or less without incident. At first, there were no signs of trouble as only the small amount of field trip decliners—5 or 6—chose to stay back. But as anyone familiar with schooling can attest, most *any* field trip tends to attract most students as a way to evade school for the day. The intimation that there might be tension within the group came as some department faculty sensed negative feedback for their program. At the end of the day, "some of the kids left saying that 'they weren't from monkeys'" as Dr. Russell recalled. As he continued reflecting on Darwin Day: "When I was teaching on Darwin Day…there were chaperones that would hide their face, kind of like "I can't hear this, I'm not hearing this…I can't really be exposed to this"…it blew my mind."

Dr. Russell was the only faculty member to recall such intense passive resistance to Darwin Day. What he saw though was simply an indicator of greater resistance to come. As events would soon unfold at Nolte County Schools, conflict was soon apparent. "A seventh grader heard about the eighth grade going to Darwin Day. This seventh grader's mother works at the Creation Museum. They got back to the principal…and the Creation Museum told them that they wanted equal time." Dr. Russell was incredulous as he retold the story. Describing how he checked on the administrative decision-making process of this with other local school administrators: "A former administrator was like…'that's got to be the school's decision and we can't barge in here'…and I was like wait a second!." Dr. Russell paused, almost crestfallen. "So apparently, they gave them an hour....the Creation Museum got an hour." Keeping myself in check given the legal implications of this move by Nolte County, I confirmed that this was a talk to their science classes. The principal of the school, as Dr. Russell recalled, allowed them in. Recovering from the gravity of what, by lay eyes looked to so clearly violate federal law, Dr. Russell rescued some consolation through his discussions with Nolte County science teachers:

> Apparently, after the talk the teacher got all the students together and asked them, given what they saw at Darwin Day, which one makes sense? He said that they saw right through the Creation Center and didn't really buy into the spiel that they were trying to float.

Naturally, I wondered what had come of the Answers in Genesis visit. Practically begging for it to be a news story, Dr. Russell oddly deflected. "If I lived in this state…I can't believe someone hasn't contacted the ACLU about this." Russell was a resident of an adjoining state which seemed, in his mind to relieve him of any responsibility in bringing such events to light. In his own earlier discussions, faculty at a prior college where Dr. Russell worked "took out a full page ad" to ward off Creationist influence. As one might also presume, in this case such actions might only work to further polarize the local relations toward Mason-Dixon, perhaps even further bifurcating the public regarding evolution.

6.5.2   Earth Day

The politically charged nature of evolution in the civic discourse extended to local festivals. Gearing up for the spring Earth Day event, natural preserve proprietor Don Atkins along with faculty friends from Mason-Dixon thought to set up a table promoting evolution. This was the second year that Atkins and his friends at Mason-Dixon, having solidified their partnership through Darwin Day, would participate. As another form of outreach, this table would additionally serve the purpose of making Mason-Dixon's biology program more visible. Earth Day would take place right downtown at a premier park in New Providence, the urban core of Mason-Dixon's region. In response to their application for a table at Earth Day came a surprisingly qualified reply (again forwarded by Mason-Dixon faculty):

Mr. Atkins,

I have received your registration for the Earth Day event at Sanders Park and I called yesterday and left you a message to call me concerning the exhibit. We need a more detailed explanation of the exhibit on evolution before I can accept the application. Due to permitting restrictions at the park, we are unable to accept registrations that are political or religious in nature. We also need to be certain we are not viewed as promoting evolution versus other 'means' of creation. Please call me at my office at (555) 555-1212 so we can discuss your exhibit.

Sally Conklin
Earth Day Cochair

As a bit of investigation would find, governmental political pressure was not the root of the concern. City park officials confirmed that as a matter of constitutional protection, no group with the proper paperwork filed would be denied participation. In fact, religious groups regularly scheduled use of the park for festivals and outings. The concern came from within the Earth Day sponsors. As best as I could find, the past year had seen some concern that a homeopathic healer was exposing children to non-Christian views about faith and medicine. Regardless of the source, the Earth Day event, organized and primarily sponsored by the local Environmental Protection Agency office, was being cagey regarding who might be potentially alienated by evolution. Rather than directly address why evolution might be a problem for some, the Earth Day organizers instead shuffled blame to the faceless scapegoat of "permitting restrictions." What then was the political agenda? How had Ms. Conklin "learned" that the factuality of evolution somehow violated an interest of the Environmental Protection Agency, which they must not be seen as "promoting?"

6.5.3   2-for-1 Day

Unassociated with Mason-Dixon in any way, both the Answers in Genesis Creation Museum and the New Providence Zoo got a round of variably (un)welcome news coverage during my time in the Mason-Dixon region. Playing on the appeal of regional tourism, Answers in Genesis and the marketing department of the zoo announced a partnership by which discount coupons would be available for attending both jointly. The marketing ploy was short lived. Within 2 days, the zoo received a tremendous amount of member complaints, some from large private and corporate donors threatening withdrawal of support. A statement by the zoo public relations department illustrated the lack of understanding of what such partnering meant for the legitimacy of the institution regarding science. "When we partner with the [sports teams], we don't get these kinds of e-mails," zoo spokesman said. "It's pretty clear this is more of a distraction." This kind of relationship is precisely the kind of affiliation that scientists and donors complained about in the volume of e-mails and phone calls they placed to the zoo. As they saw it, the zoo had garnered a top-quality international reputation for its scientific research which Answers in Genesis could symbolically trade upon. In contrast, Ken Ham of Answers in Genesis took the opportunity to complain about the "intolerance" of those inclined to complain:

Certainly, this has thrust the Creation Museum…into the news. I believe, as a result, this has really taken our…promotion to a much higher level. We can certainly thank the intolerant atheists for this! These people have been ranting and raving on their websites about the issue—but as usual, using emotive, blasphemous, and vulgar language as they denigrate God, AIG, my name, etc.

What is *very* obvious from this is that those who caused this controversy are not just against AIG's straightforward reading of the Bible, they mock Christianity; they mock the Bible; in fact, they often mock anyone who believes in God regardless of what position they take on Genesis. It is an intense intolerance (in fact, hatred of) anything that is Christian in any way.

Like any press being good news, Ken Ham in the end beamed about the additional national news coverage these actions have spurred for the Museum, and its overarching interest in evangelism. Couple with this the general observations that Edgell et al. (2006) have made regarding atheists in America and the whole matter played out as a successful rhetorical ploy for Answers.

In this chapter, we have moved past our close analysis existential anxiety, ontology, and worldview change by looking at the cultural milieu of the university campus and its practices toward evolution. In the next chapter, we continue this exposition by returning home to the classrooms and communities in which the foundation for university science education is laid.

Chapter 7
Evolution Education from Campus to Home

You have paid full fair, so I'll carry you through
For it's only right you should have your due
Why, the laborer always expects his hire
So I'll land you safe in the lake of fire

Where your flesh will waste in the flames that roar
And my imps torment you forevermore

Hell Bound Train (excerpts)
American traditional

7.1 What Students Have Learned About Evolution

As I have shown, there is much "known" regarding evolution that matters deeply for education, and is well served by discussing ontology and worldview. Continuing my analysis, this two-part chapter first articulates four critical aspects of evolution education as these students experienced it. I present an overview of student perceptions of their past science education in general. Next, I discuss the ways in which students have come to know evolution as something that teachers specially qualify, and where/when evolution becomes constructed as contentious in their lives. Fourth, I discuss how Creationists have learned to discuss or silence their views, depending on who is listening, all the while scoring highly on class assessments.

The second part of this chapter turns toward interviews conducted with case study student's high school teachers. I present these teachers' perspectives as vignettes parallel to general themes about evolution that emerged from interviews with students. Each vignette is paired as an exemplar of that theme in practice, in some cases ironically. I do this with an eye on animating a picture of those discourses and cultural practices, typical of many US classrooms, which work to avoid or negate evolution from discourse. I first review three teachers' experiences who do describe teaching evolution, if in unexceptional ways. I came across few exemplary

D.E. Long, *Evolution and Religion in American Education: An Ethnography,*
Cultural Studies of Science Education 4, DOI 10.1007/978-94-007-1808-1_7,
© Springer Science+Business Media B.V. 2011

classrooms where evolution education was richly discussed, itself a matter of thematic importance. Continuing, I then discuss three additional thematic points in detail: the empirical ubiquity of a "teach all the theories" ethos expressed by students, the surprising sentiment: "I'm a Catholic, I can('t) believe in evolution," and perhaps the most pointed theme—"the *really* religious people." I close this section with two short vignettes—an exemplar of teaching excellence and a view of student indifference toward science. To end the chapter, I discuss the political dimension behind these discourses.

7.1.1 Prior Science Education in General

In interviews, students discussed their life histories with science education, religion, and evolution. Across all their experiences, student science education generally appeared pedestrian to me in that it appeared so "normal." There were few glaring holes (barring evolution) in their content preparation, and their participation in high school science included biology and dissipated after chemistry or, for just a few, physics. Three students were exceptional in that they discussed having had no elementary school science classes, each coming from schools that they described as focusing heavily on "reading and math."

The students I spoke with were far closer to traditional matriculation age than what I expected, given the quickly changing landscape of participation and form in higher education. Almost all were full-time students, although they ranged in age from 18 to 37. None had children. Almost all of them worked either part or full time on top of their university class commitments. Although I expected to find some community college transfer students, I encountered none. A few had attended other regional, city, or liberal arts colleges, but they were atypical. These few tended to have been to at least two prior institutions. Not atypical for a regional comprehensive university, three quarters of the students were first-time college goers within their families. Only a few had parents with bachelor's degrees or higher. Students described choosing Mason-Dixon for two main reasons. They perceived it as a high-quality public school education alongside quite relatively low costs.

Although now living in the region, a slight majority of students I interviewed started their lives all across the nation. California to Virginia, Minnesota to Tennessee—students had moved around in their lives trailing their parents at the whim of the job market. Of the large minority who grew up in the region, their home communities were within a 200-mile radius of Mason-Dixon, coming from four states. These students, as a group, were from all over the USA, with the exception of the American deep south.

In the early part of my interviews with students, I surveyed their experiences with school science from elementary school through their attendance of Mason-Dixon State. Walking me through their experiences, I asked them first to describe their relationship toward science, what interested them and what did not, and then what they remember of their school science classes from elementary school through high school. Unsurprisingly, some thematic commonalities emerged regarding their interests in science. About three-fourths of all students distinctly spoke of their dislike of chemistry, which was immediately clarified with the qualification "because I don't like math." Of those who had distinctive interests in chemistry, four of six were Creationists. Typical of this, Andrea explained: "I really like chemistry. I really like working with the elements and seeing how the electrons go from place to place and how that really is the basal structure for everything else that exists." Although I did not investigate this further in my interviews, this attraction to foundational knowledge squares nicely with both Wuthnow's (2005) earlier description of philosophical absolutism and Toumey's (1994) work with Creationist engineers, who also were attracted to the perceived fixity and interpretation-free outlook of these "hard" sciences. As Andrea explained further, "I like to have a lot of facts to support my own ideas. I think that learning how things work really helps you to be able to do that."

Perhaps due to evolution's conceptual undermining of her absolutist and Young Earth Creationist views, as seen in Chap. 4, Andrea rejects foundational science supportive of evolutionary theory in favor of the cataclysmic "flood geology" of the Creation Museum. Tyson shared a similar outlook on the perceived fixity of some sciences. "I like things that are very practical. I think a lot of students are like that. I'm kind of a math guy, so I like chemistry especially. It's not just memorizing. You can put numbers to it all." Although perhaps only a point of curiosity, no Creationist students singled out Chemistry as something they avoided, whereas almost all "Both/and" students gave the thumbs down to chemistry. Many students of each persuasion spoke highly of biology, if only referring to cuddlier topics such as "liking animals" or having a general interest in "finding out how the body works." As stated in Chap. 4, the atheist and agnostic students spoke the most authoritatively about the nature of science, and volunteered interest in the most conceptually advanced topics—quantum mechanics and the specifics of human paleontology.

7.1.2   Qualifying Evolution

As part of my interest in student histories with prior science education, I wanted to draw out their exposure to evolution. Given what survey research had detailed on this subject in the past, I expected to get good examples of resistance to evolution from parents, perhaps clergy, church youth groups, and in some cases teachers themselves. I also expected to find a few exemplary examples of evolution being taught. The extent to which evolution was qualified or downplayed was more pervasive than I thought it might be.

There is methodological difficultly in trying to represent student exposure to evolution in classes from 31 differing teachers, different states, and the range of public, private, and parochial schools which these students attended. Rather than presenting decontextualized measures of an enacted curriculum, I needed deep relational information about how science and religion interact in educational practices

Table 7.1 Student perception of high school coverage of evolution

Time on evolution	0	1	2	3
No of students	7	11	9	3

0 No time on topic; 1 A class period or less spent on topic; 2 A half week to a full unit on topic (unit was detailed as 2–4 weeks); 3 Evolution fully integrated across curriculum

and the significance of these in one's understanding of a world. I was also more interested in how students came to understand evolution as socially contested knowledge rather than their factual understanding of the concept. I questioned students as to their recollection of length of coverage on the topic, in what classes (it was not always a science class) evolution was taught, and asked them to recall how the teacher introduced or discussed the topic.

To conceptually represent student perceptions of raw time spent on the topic, I aggregated student answers by the time on topic that they described. Using a scale from 0 to 3 detailed in Table 7.1, student prior formal education with evolution was as follows:

Reading (roughly) across the range of student perceptions, with regard to how evolution was taught, some stunning insights are apparent. According to student recollection, for just over half the students I spoke with, evolution was either never addressed or mentioned in one or less class period—during the *entirety* of their K-12 education. One might then presume that there was a relationship between the quality of the classroom teaching and student attitude toward evolution. Based on how these reported experiences broke down by ontological position (Christian exclusivist, the Both/and, and agnostic and atheist), I saw no appreciable relation. In fact, a somewhat disturbing and deflationary relationship was evident. There were the few cases where students articulated a fairly orthodox conception of what evolution is, all the while having had no formal K-12 experience with the concept. These same students often described learning of evolution through media outlets such as the Discovery Channel, Public Television programs such as NOVA, and more than one described learning about evolution (at least the social controversy surrounding it) from the *Simpsons* and *Family Guy* animated sitcoms. Additionally, as interviews with their teachers confirmed, often Creationist students emerged from what I would judge to be exemplary high school biology classrooms. There was also the very odd case featured in a later vignette where a student clearly recalled his high school biology teacher "skipping that chapter… because it was a Catholic School" only to be contradicted by the same teacher in a later interview who "teaches evolution as an integral principle to [her] classes." As I am working toward, the vocabularies that worldview can proscribe, and the ontological categories that these make possible, have a direct impact on how one approaches science.

As stated above, the topic of evolution was qualified and entrapped in social contention within both middle- and high-school classes. Of the 31 student experiences I encountered, 12 clearly recalled one or more instances in which their science teachers or school administrators delegitimized evolution by rhetorically casting the concept as illegitimate. For example, "we're just covering for the state test... and you don't have to believe it." In seven cases, students recalled their teachers specifically holding this kind of meta-discussion, essentially devaluing any authoritative position evolution held in the curriculum. Two other students spoke about their schools sending home permission slips to secure permission "to do the evolution lesson." In these cases, evolution would be a once-off topic which students could elect to sit aside with "an alternate assignment" similar to students objecting on moral grounds to dissecting frogs. As featured in this chapter's *Litmus Test* vignette, in one extreme case, evolution was forbidden at the institutional level due to the influence and passive resistance of a majority of one campus's students, egged on by some Southern Baptist ministers in the area. In three of these twelve "qualified" classes, evolution was covered for a half week to a unit. One additional class (Tyson's homeschooling cooperative experience) did have an unusual amount of critical attention placed upon evolution, with the caveat that the entirety of this was a Creationist household guided by a Creationist homeschooling ethos.

7.1.3 Constructing Contention

Surprisingly, *how* evolution became controversial in the minds of students had a thematically common origin in student narratives. Separate from my questions of the coverage of evolution in their education and lives, I asked them where and when they first became aware that evolution is a controversial topic, if they ever had in their lives. Time and time again, often including the same teachers who held meta-dialogues prior to "the evolution class," students reported first hearing evolution as controversial from their middle school science teachers. Carly first remembers hearing this in "middle school biology...I guess that's when they first mention it. But they don't really talk about it too much because you're not supposed to teach it supposedly." When I asked her to clarify exactly who stated this, she confirmed "the science teacher."

Tanya, who attended a specialty magnet high school that consistently scores in the top ten for academic performance in her state was clear about evolution not being taught. "We don't talk about evolution...or we didn't." When I asked her to clarify why she hesitated, she explained that the topic might come up in a side discussion regarding the age of fossils or the lineage of certain animal species, but these questions always met curt responses from the teacher who added "but there's also other theories as to how we came about." Indicative of the kinds of administrative pressures I heard discussed in a few schools, James (who previously learned that evolution is the work of the Devil at home) learned that evolution is to be avoided at school:

> It was an iffy subject. It wasn't something that they were…I guess the principal, or maybe the school board told them not to delve too deep into it. It was kind of a faux pas—taboo area when I was in school at least. Coach Sims, the teacher—when he brought up evolution he was like well, you know, "but I can't go into it because they told me not to"—like the board told him not to.

Recalling Cindy's experience in her dominantly exclusivist Christian family and school setting from Chap. 5, "the teachers didn't want to go too in depth with it because they were afraid that they were hitting points that the students didn't want to hear due to religion." As we saw in Cindy's case study, her high school teacher seemed frustrated by the social pressure to not teach about evolution. After showing a short video "approved by the administration" on evolution, she sarcastically summarized, "well…we covered it!" Cindy's case was extreme but not unusual. It appears that some school discourses about science are simply muted due to the general influence of religious exclusivist worldviews and what these dictate in normative social practice.

Andrea, while not attending a school that eliminated evolution, did have teachers fully aware and practiced in how to handle certain "trouble students." As one of Andrea's friends alerted her to the impending "doom" of evolution:

> One of my friends that was in the class before mine with the same teacher and had mentioned to the teacher 'oh well!—Andrea isn't going to like this because I just don't believe in it'—So the teacher came up to me before class started and warned me, 'hey, we're going to be talking about this, but I'm not going to teach it as a fact, so you don't need to worry'. Like really making sure that I wasn't going to cause a big deal. And I'm not the kind of person who would do that. I just kind of sit back to myself.

Cindy, like Andrea, who both are preparing to become teachers, clearly saw the problems of teaching evolution cleanly in her community. When I asked her how she might achieve this:

> If you want people to learn about it, you're going to have to approach it in a way that they're not going to be offended by it. So like in my hometown (*pauses*)— I'm trying to think how you would approach it. I mean, there would be no easy way to approach it, really. If I were there teaching—this is tough. I don't know how I would even go into it.

7.1.4 That Which Remains Unspoken

I structured the approach to my interviews so as to collect data from a position outside the power dynamics of professor and student. I was free to explore areas of thought and discussion into which a faculty member might never venture, feeling it inappropriate to ask. What, for example, was going on where genetic engineering was a topic that prompted much discussion in Dr. Green's class but evolution did not? Perhaps not a point of contention for Both/and students, but certainly for Creationist, sitting through classes on evolutionary theory prompted irritation or at least a bit of alienation. In more than one case as happenstance would see it, Creationist students who had been selected as case-study participants scheduled their interviews just before evolution was scheduled on the course syllabus. Naturally, such timing allowed me a natural entrance into their anticipation about how evolution would be treated. When I asked Andrea how she deals with the inconsistency between her beliefs and what her college science courses teach, she explained:

> I usually just sit back to myself. And if something is said that I think is just really off the wall, I'll bring it up if I have a reference point to back up what I'm saying —but I'm not going to cause a big issue in a classroom.

But what an interview transcript cannot convey is the tone of irritation or bother that Andrea disclosed. Explaining why the tone of the college class bothered her:

> Every time it has been presented to me outside of this college, it has been presented to be as you know, (*mimicking the teachers in her life*) 'this is the way that the book reads it, this is what you have to know for the test, this is not what you have to believe'. That's always what I've been told—every time it has been presented.

Andrea was not a poor student by traditional measures. In the content of her advanced placement high school biology class, she described her teacher as "not wanting to linger" on evolution for the reasons detailed above. As I later discussed with this teacher, evolution actually gets quite a bit of coverage from him in the content of the class that Andrea took. Why then the discrepancy in perception versus reported practices?

In Tyson's case, the political implications and fears of discussing his views are more palpable. Having been homeschooled, Tyson had been presented evolution, but always as the product of a false, ungodly secular worldview. I was curious how he was going to deal with the fact that evolution would be taught as truth in his college class with no mention of Creation as a viable alternative, in his case for the first time in his life. Tyson detailed what he perceived to be at stake as I asked him how he would negotiate this in his mind:

> (*Hesitating and thinking a long time*). I should think about this....I have sort of thought about it...but, people do different things. Some people I know will answer like...if it's an essay question they'll write an answer and say 'Professor, this is what I think you want, but this is not what I believe', and they'll write what they believe over here. So they'll actually write two essays...and that's a freaking lot of work and I don't know if I'm going to be doing that. So if he asks if like...prokaryotes.. did they evolve from a form of mitochondria, I'll mark that...I'll say yes they did...I don't believe that, but that's what he's looking for. As of right now, that's what I've come up against. I should talk to people about that.. I want to figure out what God wants me to do on that—you know what's the *right thing* to do.

Given Wuthnow's (2005) earlier insights about exclusivist Christians being limited by the restricted social networks of the insular congregations, this is not surprising. As Tyson discussed in Chap. 5 how he conceived of other worldviews, his view is oppositional, Godly versus those who are not. There has been precious little opportunity for other narrative to intercede in his life. Andrea expressed similar conceptions. As her religious commitments are very similar, her story is doubly interesting as she is also planning on being a secondary science teacher. Andrea explained her anxiety at entering this field: "I feel that in my field that my views are not usually accepted. I feel that people who might have the same aspirations that I do might feel (*long pause*) a lot of pressure...." I asked where this pressure comes from. "From the world. Teaching subject matter such as this to the point that they won't even touch it—because it is such a controversial thing that—they might not pursue what they want to do." Given this contention and Andrea's future plans to be a science educator, we discussed how Andrea will treat evolution in her classes:

> I've thought a lot about this. When approaching the subject I'd really just want to present the ideas that are given and challenge my students to put together the facts for themselves and see where they come out with believing it. I'm asking them to challenge their own minds and believe what they believe because they have a basis for it. And I don't care what they come out believing as long as they have their own basis for it.

Still searching for an ultimate foundation or *basis* as she repeats, Andrea appeared caught between what counts as fact by the overwhelming coherence of scientific consensus, and the contradictory factual direction of her Creation narrative. Illustrating this semantic difficulty, Andrea detailed her quest for final unchanging *Truth* as she conceives her philosophy of science:

> A theory is something that cannot be fully tested. It's something that we cannot know the extent to its factual basis—gravity is a theory—it is not a fact. It's not something that we cannot fully grasp and test and see and be able to work with. It's something that's assumed.

It seems Andrea's confusion as to the status of formal theory solely results from her culturally imbued absolutist epistemology. As such, her ontology precludes her from seeing any other middle course or, even more bothersome, the possibility of an ungrounded interpretive understanding of science and religion. In similar form to Tyson, Andrea both understands little other than what her small circle of religious thought has fostered, as her reflective discussion illustrates. When I asked her to discuss other worldviews than her own, she also struck an immediate oppositional stance:

> As opposed to a Biblical worldview?—there's basically a scientific worldview where everything is based on what's in a journal. I think there are also a lot of spiritual worldviews which would be different from the Biblical worldview in the sense that they're more about how you feel about everything and kind of seeing everything through rose colored glasses.... (*very long pause*)
>
> I don't know—it's not something I think about a lot—because through my whole life there's been like what everybody else thinks and like what I think. It's always how that feels...it's like one or the other. So, I've never really thought about the multiple ideals that are out there.

This type of limited thought regarding the ecology of one's faith amidst others was already clear enough in Chaps. 4 and 5, but is worth remembering as Andrea is very close to likely entering a high school science classroom. Teachers, like students, have religious dispositions, and they are not always supportive of evolution.

7.2 The Teaching of Evolution in Schools

In this chapter's second part, I turn to vignettes that illustrate high school teacher experiences. I present these vignettes in parallel to general themes in students' sociocultural imagination about evolution that complicate evolution education. Each vignette serves as an exemplar of that complicating factor in practice.

The contention that Darwin Day prompted quickly reminded me of what I might find once out in schools talking to teachers. As planned, I spoke with those students' teachers whom I had selected as case-study students represented in Chap. 4. As it would happen, there was little or no relationship between the time high school teachers spent discussing evolution and the students' receptivity toward it. Some students who described fairly orthodox or even nuanced understandings of evolution came from classrooms which omitted the topic. Over half the Creationists I spoke with came from classes which at least briefly discussed the topic. Student recollection itself might not match with teacher practices as one case I will share will make abundantly clear. As we know, from my earlier discussion in Chap. 2, teachers omit, gloss, or teach Creationism in parallel with evolution in many cases.

So what is going on in American high school science classes regarding evolution? Of the 12 case study students selected, I managed to meet with eight of their teachers. The four missing were either referred from a nontraditional age student who could not recollect their teacher, one who had changed fields and could not be contacted, one who was known as deceased, and one who was retired and simply could not be found. Of the eight teachers interviewed, their work experiences with schools and the treatments of evolution within opened a sort of randomly selected spiderweb of social and institutional scenarios. In this way, these eight introduced me to over 12 schools worth of evolution education.

Given that case-study students were selected for their religious ontology, the collection of teacher experiences I share were quite random. This is not to imply that I intend readers to see correlations or generalizations. What I do intend is for readers to keep in mind how easily and deeply I found antievolution shenanigans afoot in schools. Although standardized testing may never show such things, and as students' perceptions of the social construction of conflict in schools strongly alluded to on whole, there is a great leveling down of discourse regarding evolution in our schools. In most cases, this is certainly to avoid social conflict. Talking with Creationists about evolution requires talking about how one conceives of and operationalizes views of reality. Such discourses do not respect the intellectual boundaries that some see science and religion quartered into. Similar to the students themselves, there was no strong reason to presume great theological literacy of teachers. Like anyone else in the kind of sociology of religious life that folks like Wuthnow (2005) describe, teachers have religious commitments too.

Before students arrived at Mason-Dixon, the types of school settings I will detail were the political environments in which evolution was presented (Table 7.2). Some appeared excellent, some mundane, some infuriating, and some possibly breaking federal law. The following stories have simply had names and places changed to maintain anonymity. Within are things that made for jaw-dropping interviews, and given the rationale for this study, partly left my angry. Leaving these vignettes, I close with one exemplary example of evolution education done right—with the cautionary note that the exemplary status is garnered for simply doing what policy and curricular writers intend as best practices.

Table 7.2 Teaching evolution: map of discussion

Theme		Vignette	Corresponding student
Student social imaginations regarding evolution (themes) and their exemplars in practice (vignettes)	Positive but unexceptional teaching of evolution	Mr. Gardiner	*Andrea*
		Mr. Davidson	*Renee*
		Mrs. Hunter	*Nolan*
	"Teach all the theories"	Mr. Woodman— "what about in social studies?"	*Lauren*
	"I'm a Catholic, I can(t) believe in evolution"	Mrs. Weidman— "you better watch out"!	*Chad*
	"The really religious people"	Mrs. Garcia—litmus test	*Tonya*
		Mrs. Petit—the evils of secular humanism.	*Tyson*

7.3 Student Social Imaginations Regarding Evolution and Their Exemplars in Practice

7.3.1 Positive but Unexceptional Teaching of Evolution

As Cindy's story already showed, evolution was contentious within her school. Owing to the dominance of exclusivist Christian practices, both the classroom and the school context in general made evolution a dangerous topic. Students rebelling by not coming to class, administration overseeing "approved" videos concerning evolution, and a general climate hostile to the concept punctuated Cindy's experience. Having left the teaching profession, her teacher was in part frustrated by this climate by Cindy's assessment. As already discussed in Chap. 5, James' recollection of evolution in class was similar in that the district administration forbode the teaching of the concept, at least as James recalled it.

When I spoke with Andrea's advanced placement biology teacher, he explained his focus on evolution which he described as being "all throughout" the AP course. I did not get the impression that Mr. Gardiner was intentionally avoiding evolution as Andrea had perceived. Both Andrea and Dr. Green at Mason-Dixon during her school years had taken his advanced classes. Each spoke highly of the experience. Gardiner did give the impression of a teacher extended to the limits of time and commitments. So why the difference in perception between the enacted curriculum and how Andrea perceived it? As I argue in Chaps. 4 and 8, ideology certainly plays a part in mediating perception. The rhetoric of a high school class, the nature of its compulsory clientele, and the socialization of high school teachers versus college faculty likely played a bit into this difference. Whereas Mr. Gardiner "didn't care what you believed…that's for you in your own mind…this was science," he was in ways similar in approach to some Mason-Dixon faculty. But Gardiner also makes the slight rhetorical accommodation toward areas of knowledge respective of Creationist students which few college faculty would likely make—"I don't care if you don't believe this, as long as you at least listen to what I have to say." Although not common, he was aware that Creationist students would simply tune out if alienated, putting their heads down on the desk if they did not care for the narrative. Recalling earlier experiences where he would not predicate the evolution lessons with a nature of science discussion, Gardiner recalled that students often pejoratively turned on each other. Comments such as "stupid, uneducated, Satanic, godless, evil" typified these exchanges. Although he chalks it up to lack of time, Gardiner currently avoids discussing human origins or the implications of geologic time.

Within his school culture, Gardiner saw some social conflict regarding evolution. He expressed that in prior years he worked with a Creationist science faculty colleague "who completely rejected evolution, but taught it in his class…he walked a very fine line, and I don't know how he did it." Gardiner explained that this teacher's epistemological balancing act always confused him. As he explained to me, he also infrequently encounters resistance from other faculty when he teaches (*using ominous tone*) "the evolution lesson." Dr. Gardiner also felt that the regional influence of the Creation Museum also had a tangible effect on the editorial board of New Providence's major newspaper. Echoing sentiment that Mason-Dixon faculty shared but could not substantiate, Gardiner explained that he felt that "they must have someone on at the paper…every time there's a news article about fossils or evolution…I always look at the editorial page." As Gardiner explained, with a sentiment similar to one that Dr. Russell had shared with me, the New Providence paper was disproportionately active in allocating column inches for antievolution stories on their editorial page.

Mr. Davidson teaches high school biology just a short drive down the road from Mr. Gardiner. Like Gardiner, Davidson presents a fairly robust unit on evolution to his students. As he sees it, his school curriculum and his teaching are highly aligned with state curricular content that gets assessed on standardized tests at the end of each year. "That things change over time…it's a big part of the state testing. We do 5–6 weeks on evolution. The history of Darwin, Lamarck, Wallace… their travels." Along with the mechanics of evolution, Davidson also mentions that he plainly covers human evolution, the first high school teacher to mention to me that they do so. Although not a dominant phenomenon in his experience, Davidson explained the kinds of resistance he encounters in his teaching. Learning from his new colleagues as a new teacher, he was warned to treat evolution gingerly. They explained to watch out, "that conservatives and fundamentalist students and their parents could make a scene." I probed as to what constituted "a scene." As he explained, it would be their speech. "That it was evil. Some of the students were withdrawn by parents from the classroom." In light of this climate, Davidson explained what he feels successful and problematic in teaching evolution:

> If you use different terms—if you don't use the word evolution—you get a lot of mileage. If you talk about sequential adaptation—change over time—they're very open to that. When you say evolution, that's such a charged word for them, they don't want to hear that.

As he explained, this type of rhetorical footwork is most sensitive concerning human evolution. "Those people who even are highly opposed to it, until [human evolution] they seem to see that it works." As seen before with other students, and echoing the concern of the epistemological foundation of exclusivist Christians, the mechanisms for evolution work "until you get to humans, and then they block it." The issue is clear enough regarding the primacy of a special creation. "They'll say I understand that to that point." Davidson recalled the clearest example of how this issue suddenly tips in class discourse. After discussing the evolution of humans, "I had one woman say I didn't know that that you don't believe in God." This followed with an exchange echoing that which came of Dr. Wayne's Pascalian bargaining in class. Challenged by a student as to whether one can be a Christian and believe human evolution had taken place, Davidson responded "well I do" met with a blunt "you can't." As he animated, this goes on back and forth for a bit in the classroom.

Issues regarding class assessment that Tyson discussed earlier also surface for Davidson. "In advanced placement testing, there's always an open ended question on evolution. A student wrote in big letters, I believe in God...down with evolution... that was her response." Although extreme, Davidson rarely saw this. But as seen in Tyson's internal dialogue, such behavior, while hard to catch, illustrates the rationales of many students in similar circumstances. As one Creationist student Davidson recalled sized it up, "I know it's part of the science curriculum, I know I need to know it...but I don't personally believe that."

Similar by amount of coverage on evolution, but with a differing rhetorical tact, Mrs. Hunter started right off with concern from parents. "Parents approach me at the very beginning of the school year." As she explained to them her approach: "It's more a genetics-based and how things have adapted to their environment approach. More than a 'I used to be an ape swinging in a tree' approach as I tell them...we don't even go down that road." Mrs. Hunter's lesson focus contains more of the mechanics of genetic drift and allele frequencies, and far less (or no) historicizing common to Davidson's classes. Seemingly prepping for potential conflict, Mrs. Hunter was judicious in her rhetoric:

> When I begin to teach it I'm very open with the kids. The very first thing I tell them is that I won't tell them what I believe. I explain to them that I don't tell them what I believe because I don't want them to feel like I'm influencing them one way or the other. I want them to be open minded. And of course, they all assume that because I teach biology that I believe one way. And that's when I inform them that I go to church ...and that's when they're like...oh, okay.

Open, closed, open, closed. Although maddeningly contradictory at times, Mrs. Hunter is indicative of a position we have examined before—the Christian inclusivist. For her students, the punch line resolves with her clarifying her position, which is clearly not "scientist as atheist." Although she attends an evangelical church that takes an antievolutionary stand, she sees herself as unique within this community. This position is typical for her family although not completely. Her brother, a "very religious" Catholic as she described him has been decisive with her regarding her views on evolution. "He does not believe that humans evolved from apes." Hunter's class and adjoining laboratory facilities strike me as extraordinary for a high school. She explained that she is blessed as a science teacher with two Principals who themselves were science teachers, and thus she has received nothing but support over the years for teaching evolution. This did not extend to the rest of the faculty. On occasion, the following kind of scene would play out. Typical of a general sense of negativity toward evolution by some faculty: "I had an English teacher approach me. 'I heard you are doing evolution in your class'…and I said 'you have to—it's part of the state curriculum.'" As she continues this teacher's complaint "I don't know why we're teaching that or why we have to."

7.3.2 Teach "All the Theories"

Apart from questions that illustrated students' past experience with evolution, I was also interested in how they felt their own view agreed or differed with what they imagined regarding the norms of the social world around them. As an open-ended wrap to my interviews with students, I asked them what they thought was the best way to teach about evolution. Students once again differentiated their answers based on the ontological positions of Christian exclusivist, the Both/and, and agnostic/atheist. As in my Chap. 4 discussion of how this made the Both/and student position distinctive, Both/and students thought "all the theories" should be taught. They resisted ruling out other voices and their "theories," drawing no hard line as to what "theories" fell within the methodological orthodoxy of science. Of 31 students asked, only five objected to including "all the theories" in educational practice. As stated earlier, four of five atheist/agnostic students asked this question clearly objected. As Sheena summed up this position regarding teaching evolution:

> I would bring it up more freely. I'd start in elementary school and just bring it up. I wouldn't worry about stepping on people's toes, I wouldn't worry about offending people's religion. It's a scientific theory…it should be free to be taught without worrying.

The fifth, quite a discrepant case for this and many other ways, is discussed at the end of this chapter.

Given the social contention that Creationists encounter, some Creationist students did seem keenly aware of what teaching "all the theories" meant politically. In his reaction to the premise that Creationism or Intelligent Design might be taught alongside evolution in public school science classes, Tyson beamed. "I'd be completely shocked, but I'd say way to go for those teachers…that's hardcore." Tyson quickly thought about the implication of what he had just said and reconsidered:

> As a teacher, you're trying to get to the *Truth*. That would be very difficult. I wouldn't believe [evolution] is true based on rational reasons. If I can only teach evolution, I'd probably quit my job. I couldn't do that. I couldn't lie to my students. I don't think anyone should do that. I'd try to find a way I could teach at least Intelligent Design. I think—I'd try to get both.

The proposition of evolution being lies, or lacking a factual basis, is seconded by another Creationist student. Andrea, who also might soon be facing this issue practically as the sole secondary science education major of my interview pool, explained:

> I would make it more just…I want to say facts based. I feel like when teachers teach on a subject, they really put a lot of their own ideas into what they're telling you. I think that it really should be taught like (*tone of emphasis*) *this is the evidences that have been given, these are the ideas that have been presented,* and *these are their misgivings,* and let the student put it all together and figure out what they want from it instead of saying this is how it is. It should be what do you think about the evidence that has been presented.

Interestingly, Andrea and Tyson both make appeals to "facts" and *Truth* in their interviews. For both, and similarly to many of the other Creationist students, when evolutionary theory by the mass of its evidentiary coherence becomes as factual as any scientific concept can be, it can no longer be a form of *Truth*. For these students, it seems this judgment is based solely on a perception that evolution undermines the commitments of their religious narrative.

7.3.2.1 Vignette: What About in Social Studies?

It has often been suggested that discussions of Creationism and Intelligent Design might be appropriately discussed within the context of a social studies class. Like the study of social movements generally, the appropriate historical context might be an effective way in which everyone can learn the stakes and terms of this cultural "debate" without taking time from content of science classes, and in this way, everyone could learn "all the theories." Naturally, I was excited when I had learned that Laurel had gone to high school under such an arrangement. I did still find it curious that Mr. Woodman, her world religions teacher, had taught evolution while the biology teacher had skipped it entirely. With the benefit of the doubt, I arranged to speak with Mr. Woodman.

As a matter of happenstance, Mr. Woodman was a Mason-Dixon alumnus. Beginning to detail his own educational history, his interest in history, as he sees it centers on the concreteness of the subject. Contrasting this, Mr. Woodman described having little interest in most of the sciences, but with a somewhat atypical reason I nonetheless find interesting. He was not interested in "things that were more theoretical. Studying theory, or understanding—just trying to get a grip on things that were more existential—I have no interest in that at all."

Woodman left little room for doubt regarding his feelings toward evolution as he walked me through his early life in religion and in understanding science from school. In his case, the discourse overlapped quite a bit:

> It was alluded to that there's a school of thought that says we started out as primates and we've evolved. And you would see the typical posters that would be associated with that, where you evolved from primates to Homo Sapiens. And that thought kind of struck me as odd—so am I a monkey?—is what I thought. But I don't *feel* like a monkey. I don't *act* like a monkey.

As Woodman recalled it, his teacher clearly thought that evolution was part of a world "out there" which his classmates were not part of. In his faith practice, this particular internal discourse of "out there in the world" was clearly supported. When I asked him to discuss whether evolution was ever a topic at church while growing up, he explained that his Sunday school and main service occasionally explicitly taught about evolution—to discredit it.

Moving on to his experience teaching world religions as a social studies teacher, I was now very interested in how he would handle this. "It was an overview course to provide, not an in-depth, but a superficial coverage of many different religious beliefs, practices, customs, and trainings." The rationale for the course was "so that students could get an understanding of how different religions are in some ways similar and in many ways different." So far so good. Perhaps this was the sole example of an exclusivist Christian I would encounter who might be teetering on a more inclusivist view. As his description continued, the educational direction for "world religions" took a distinctly odd turn:

> What I wanted to do in terms of comparing all, was for the curriculum to be set up as a broad overview. I wanted all the religions underneath that umbrella to always apply the concepts of Intelligent Design versus evolution. That way the kids could say, 'Okay, we study this. We study Catholicism. Now where does that fall under here? How would Catholicism discount evolution'? And then I wanted them to figure out or debate—now that you know a little background about Catholicism—or we would do Judaism or Hindu, or whatever. Then I would have to break them up and go, 'Okay, now you're gonna take on the role of —you are now a believer of this. And *you* are a believer, we talked about, of evolution. We're gonna debate this'.

By the oddest and most comprehensive means that I had ever heard evolution being critiqued, Mr. Woodman's strategy won the prize. He was not only compromising evolution in his social studies class, he was being downright "mutlicultural" in doing it! Not quite sure that I had heard him correctly, I made sure we were on the same page—that he was using comparative religions, one by one, to generally critique evolutionary theory. With a prideful gusto of educational achievement (*I think I may have "passed" as sympathetic to his cause*):

> I would always come back to it. And near the end of each unit it would become— 'okay, now we'll compare religion to religion'. And then I'd say, 'Okay, now let's take it to the biological evolution standpoint. Now that you know—we've done a whole unit on Hinduism. Now let's compare that to the evolution theory. Where are the main sticking points, where do they not jive'? And I think that's where I try to tack—everything always went back to those two things, of intelligent design versus evolution.

As Woodman saw it, his community dynamics were dominated by kids who were either brought up as very religious, or "they're more of a non-believer, agnostic or atheist, that kind of approach—maybe not being atheist, but that mentality." I sensed echoes of Edgell et al. (2006) work on public receptivity toward atheists.

We got on to discussing how successful he thought this approach was. As he thought that few students might independently care to study evolution, he described his approach as it had an air of social benevolence. "I wanted to make sure that, in order to be a well-rounded student, you better know all of them." Going for what he appeared to associate with an evolutionary view:

> You better know what the atheist view is, the agnostic view, then there's also this sect over here who says, it's none of that. It's strictly biological evolution. It's why we're here, and why we look like we look, and so on and so forth.

The effect of this on some students was not lost on Woodman. When I asked if this discourse created conflict for some:

> Oh yes. I had a few kids who were very much, in their upbringing, very religious, and were a little offended by the fact that that was even discussed. Their debate would go from lively discussions or they would very much put the time and effort into their synopsis or their briefing for the debate. But when it started to get a little mushy—because good points were being presented on both sides—they would shut down, because they felt like their religious belief was being called into question. So they didn't like that. One student said that she felt like her belief—and she's Christian—so her belief in Jesus Christ became called into question, and she felt that that was inappropriate in the classroom.

So maybe all was not lost on Woodman's efforts. Despite the very fine and certainly questionable line he was walking, and the almost certainly religiously inspired motivation to use the class as a venue to do a wholesale critique of evolutionary theory, I still could sense merit in small parts of his approach. There at least appeared to be student investment by study and preparation.

Adding to the debate format he focused on, Woodman would bring in guest speakers. But now tacking backward, his guest list included a local Rabbi (a "friend of the family") who would speak in favor of Intelligent Design, and a Christian pastor who also critiqued evolution. Lest there be any confusion regarding Woodman's feelings for evolution, the worldview section of the interview crystallized matters:

> The teachings of my religion would definitely convince me that evolution is not possible. The religious teachings I have would say that if man came from a primate, that the intelligent design piece of it, the spiritual side of religious teaching, it's possible that primates started first, but we didn't evolve from primates. Primates might have been the first attempt by some superior being of intellectual design—'make it, okay, that's not exactly what I want, scrub that, destroy that, try it again, until I get it right'.

Woodman had grown up in a Christian exclusivist community—"nondenominational" as he put it—although the school he now worked in was in a heavily Catholic-dominated part of New Providence. We discussed whether evolution would be controversial in the community on whole, to which he agreed, and had actually seen many arguments over it in social settings outside school. In closing, I wanted to check on perhaps the most obvious lacunae—the omission of evolution from their high school biology classes as Laurel reported. How did they justify this?

> The Board of Education, when they wrote that curriculum, adopted it with the ideal that— we don't want to make it a controversial course. We want to leave the controversy out into the public. Let's not bring it into the building. So they just omitted it.

7.3.3   I'm a Catholic: I Can(t) Believe in Evolution

My interview questions disclosed important issues pertinent to the broad discourses of science, religion, and public understanding. The students I spoke with were almost completely bereft of a historical and sociological understanding of the structure and genealogy of the Christian tradition, as this section will illustrate regarding Roman Catholicism, and as the next section regarding Protestant students. When Julie described "Christians and Catholics and those kinds of denominations" as being potentially against the teaching of evolution, the irony and naïve offensiveness of her position is lost on her. Although religious critics like Bloom (1992) would not be surprised, the picture from my students' worlds might actually be less articulate than even he claims. Bloom makes a compelling argument as to how the nature of two distinctly American faith traditions (Latter Day Saints and Southern Baptists) encourage what he describes as a neo-Gnosticism, in which the lay Christian experiences the divine and speaks authoritatively using a limited deck of tropic Bible quotes which one situationally plays. In ways, this correlates with Wuthnow's (2009) assessment of how "Both/and" religious inclusivists negotiate the contradictory influences and demands of both science and religion on their lives. But are Americans really this limited regarding the history and breadth of our religious practices? The point of this ethnography was not to expose a seeming void in the American sociological imagination regarding our knowledge of each other's faiths or even our own. But across the interview set, almost no student articulated denominational boundaries. In addition to the trend of evangelical Christian students claiming the "true" Christianity as theirs, "Both/and" students fumbled through any attempt to articulate this cultural landscape. Often, the best I could elucidate from students was a grab bag of the major monotheistic distinctions, with some of the major historical divisions articulated. "Just religious groups like Jews or Catholics and Christian faiths that believe in like a God and monotheistic religions, I guess."

Why is this limited ability to articulate religious denominational knowledge important? Major religious figures such as Pope John Paul II have gone to the extent of publishing the Catholic church's doctrinal position regarding evolution (Roman Catholic Church 1996) within which the prior Papal opinion toward evolution is traced. In this, John Paul II accords the teachings of the Catholic Church as in union with the mechanisms of evolution by means of natural selection. So when Mitch stated in an earlier chapter that evolution "is not a popular topic for Catholics," what could he mean? Had the Pope's message not reached his parish? Mandy, a pleasant enough Catholic Creationist interviewee, plainly stated her understanding of evolution and the Roman Catholic Church. "Evolution to my knowledge does conflict with it....human evolution being the biggest part of it." Mary, an education major who we met in Chap. 4, who joined the Catholic faith upon her marriage, explained her views on evolution: "It's not necessarily something I believe wholeheartedly in, but I believe in teaching scientific theory as it should be—people should be exposed to it. It's a big part of science." Although at first she seemed to have a reasonable position toward evolution, she excused herself from any responsibility toward teaching it with a strange outlook on

Catholic school education. "I want to teach at a Catholic school, so I don't worry about that." Although an apostate Catholic who is still a theist of the loosest sense, Chad shared the tension he imagines evolution causing if he brought it up with his Catholic father. "My Dad is Catholic as can be...if I would say something about evolution I know he'd say I'm wrong." Again and again in my interviews, students made assertions of conditional extremism: "my parents are *really* Catholic." Perhaps looking toward denominational identification is perhaps an extremely limited starting point.

When so many of the students, likely indicative of a large segment of the population with only the thinnest veneer of theological literacy, hold vacuous religious knowledge vis-à-vis science, they have poor grounds by which to comment on the perceived erosive effects of evolutionary theory on faith. Demonstrative of problems related to this, a Pew Forum for Religion & Public Life (2010) survey of religious understanding found that agnostics and atheists actually outscored the faithful in understanding world religion. Naturally for Creationists, fears over evolution are real as their Biblical literalism is an antiquated epistemological view of a bygone era. But therein lies the rub and the conundrum. Religious fundamentalism is part of modernity, and the sociocultural arrangements that produce more Creationists are alive and well. As some see it, even growing. But since the critique of evolution comes from Creationists, and is tolerated by the largest part of society represented by the "Both/and" theological ontology, the loudest pushback on Creationists is almost certainly then to come from the atheist/agnostic position, which is exactly what we find today. Fighting antievolutionary epistemologies is a core motivation of the "new Atheists" movement of Dawkins (2006), Dennett (2006), Hitchens (2007), and Harris (2006). Any major swing in public sentiment toward evolution will necessarily involve the proactive steps of many more "Both/and" theists.

An additional factor to consider is the already described theological outlook of natural scientists that Wuthnow (2009) discusses. With so many scientists having either no religious commitment or "unconventional" views, joined with what Edgell et al. (2006) have described as public sentiment toward atheists in society, it is unsurprising that many symbolically associate science with atheism. As an advocate for science, I grow weary hearing students fumble through their own theological commitments, especially when this limited vocabulary is the toolkit by which anti-evolutionary sentiment is shielded. In the everyday function of schooling, it is these vocabularies and epistemological commitments of a stilted religious understanding which limit a full understanding of evolutionary theory. We are right to begin thinking seriously about what exactly we are protecting when both exclusivist *and* inclusivist religious commitments are not analyzed for what they, in tandem inaction, portend for science literacy. The following vignette considers a case where various religious identities are contesting for the ideological course of evolution education.

7.3.3.1 Vignette: You Better Watch Out!

I met Mrs. Weidman with incorrect presumptions. Chad's past high school teacher, whom he was sure had "skipped the chapter on evolution" turned out to be quite the stellar advocate for evolution. Two years into leave from teaching while raising her children, Mrs. Weidman detailed the support for evolution at the Catholic high school which she herself attended and now teaches. Within the faculty and administration as she recalls, "it was pretty much accepted. The religion department enforced it." But within this dominant discourse, there were still the objectors:

> We had a couple of teachers who—I can remember one in particular was an English teacher, and she—you'd always hear through the students. They'd say, 'Mrs. Roberts told us that you shouldn't be teaching that, and you shouldn't be teaching that in this school'.

Interestingly, although teaching at this school supportive of teaching evolution, this particular dissenter was not Catholic. "I think she was Baptist, actually."

Although evolution education was supported, along with a handful of faculty, students were at times uncomfortable with the topic. To quell this general discomfort, Mrs. Weidman's approach was comprehensive and philosophically sound:

> I'd directly address everything about it. We talked—I always had them read articles on intelligent design and try and make them think about why is this a problem, why is it not science. I can remember, actually, there was a *Scientific American* article that was the 20 best arguments against intelligent design. I can remember giving that to the kids and saying, 'Let's read this. Let's talk about it'. So I really wanted to paint it in this controversial-type light, because they were sheltered kids for the most part.

As she explained, due to her perception that many of the students were culturally sheltered, she wanted to bring them onto the same page regarding what science could and could not do, and also ground them in the idea that due to differing religious interpretation, evolution prompted controversy in society. All seemed to go mostly well for Weidman until, like Davidson's experience, human evolution was broached. Referring to a lesson examining phylogenetic trees:

> It seemed like they would be okay with all the other evolution stuff. And then all of a sudden we'd get to humans and I'd say, 'Okay, here's where the old world monkeys branch off, and the new world, etc. And then somewhere over here we have the hominids'. And then they'd say, 'Well, wait a minute. This doesn't make any sense. I thought you said we came from monkeys'. And I'd say, 'No, we have a common ancestor'. And then other kids would say, 'Well, that's not true. This is all wrong. We don't have any evidence of that. Look, you've got a question mark there in the timeline. How do we know that this is really the length of time that you said'?

The conceptual difficulty that Weidman encountered is not unusual for any classroom that teaches about evolution. Teleology and perceptual expectations of a "complete" fossil record all typify these kinds of teaching scenarios. Add to this the dimensional input of religion, and you begin to see once again how denominational identity does not quite disclose the same impression as actual class discourse:

> I was always amazed at—I would always put a question on, like an essay, about what is the Catholic church's belief, because that was one thing that I was trying to—being in a Catholic school—I was trying to drive home. And I would still say, even after talking about it for a month, maybe 20 percent of the kids would say, 'We're against it. It's wrong. Blah-blah-blah'. And they just had completed the unit on evolution. I wasn't asking 'what is your concept'. I was saying, what is the Catholic church's idea of it. And it was still a lot of kids that I clearly had not gotten the message. So that was always a little bit tough for me.

Like many educators, Mrs. Weidman had experience teaching at more than one school during her career. Instructively, little changes such as administrative support and which religious practice dominated the local community dictated much of what counted as "controversial":

> I did teach for one year at Hamilton High School, which is a public school about 40 miles east of here. And I was not teaching biology, but general science. Evolution was not part of the curriculum that I was supposed to teach, but students would bring it up. And that I found, as far as what the school told me I was supposed to teach versus what I wanted to teach—I had some conflict there. They would talk about how you have to give equal time to—I guess Intelligent Design at the time was the big thing. 'You have to talk about Intelligent Design, a little bit of Creationism, and evolution as well'.

Curious where this message came from I asked her to further detail this experience. "The science department head would tell me that. And he actually, I think, struggled with evolution as well." Although this department chair taught all of the biology classes, the discourse regarding evolution did stay within those biology classes. "I got lots of questions from students. I would have some students, depending on what I said say, 'Well I'm gonna tell my mom what you said, and you're gonna be fired'."

As Mrs. Weidman discussed with me, the very few students who would inquire for more information regarding evolution did so in confidence. "It was very clear that the rest of the class were not the kind of people who would want to know anything about that, and were upset." As she saw it, this dominant discourse clearly dictated the kinds of questions one asked. Curious as to whether Weidman had reflected on her experience in these different types of environments for teaching science, she shared with me her thoughts:

> I would say the biggest difference was the diversity of—or at least compared to what I was used to—the diversity of the kids that were coming there. They were all very rural kids, although the ones that lived in town were different than the ones that lived on the farms. But it seemed like, coming from a Catholic school education, I just kind of had this idea that everybody thinks this way, and okay, there may be a few people out there that don't. And then when I got there, it was other teachers, it was the students, and it was more of a religiously-based fundamentalist-type belief. Whereas I had been raised from more—look at the history of things, and take everything with a grain of salt—not such a literal translation of the Bible, and things like that. Whereas out there it seemed like, regardless of what denomination, most of the students had a very literal understanding of the Bible.

At Hamilton High, the tense feelings regarding evolution were palpable within the faculty in general (whereas for both Mr. Gardiner and Mr. Davidson, this was a minority feeling at their schools):

> They would talk about it at lunch, or occasionally it would come up—especially, I wasn't actually teaching that class, but in the lunchroom, they'd say, 'Oh, you're doing that chapter on evolution again. You'd better watch out. Just be careful what you say'. I just felt like there was—I can't remember any discussions in particular, but it was this sense of—well, we all *around here* know that really evolution is just something we have to teach because the state says so.

One last insight from Mrs. Weidman began to move this school and rural community in my imagination from an environment that was hostile to evolution, to one of an almost Dayton, Tennessee circa Scopes Trial caricature. "I know there were books and stuff for a while that they were talking about—you couldn't read Darwin and stuff like that. But nothing ever came of it." Incredulously, my mind began to picture the banning or burning of books.

> The public library in town for a while didn't want to carry *On the Origins of Species* and things like that. They do carry it, but when I was teaching at the school, there was a woman who was friends with the librarian pushing to remove books, so I know there was some controversy within the library

7.3.4   *The Really Religious People*

In another line of questioning prompting students' sociological imagination regarding evolution, I asked: what groups out in American society have a problem with evolutionary theory? Within this set of responses, unlike the prior accommodation of "all the theories" by at least Creationist and Both/and students, the outlook distinctly shifted. The Both/and and atheist/agnostic positions spoke almost in unison regarding the *really* religious people. As an interviewer, these responses were striking for their thematic consensus. Typifying these responses: "Extremely religious people"… "gung-ho Christians"… "the religious right folks, the hard lined Christians, Midwest and Southern people"… "the far-right Republican nuts"…and as came up no less than five times "the *really* religious people." Although many students refrained from explaining this, some had demographic imaginations. As James explained, "you know, it's kind of like, in my opinion, you got the coast, and then you got the interior. And those are—people in the interior are the more hard-lined people." Nolan was specific about denominations. "Church groups, probably. Probably Southern Baptists namely."

The distinction of "extremist" did not stop at the categorical boundaries of the Creationist. By my own categorization vis-à-vis my interest in evolution education, I categorize this position as theologically extremist, for its insistence on one final vocabulary of *Truth*. But in Chap. 5, Renee, who recently moved from secular life to practicing at a Creationist church, does not have the same categorical outlook. For her, there are additional inter-exclusivist distinctions to be made. When I asked her who has issue with evolution, she responded, "mostly like…especially the more extremist religious groups. I guess the denominational church and the ones that don't cut their hair and won't go and seek medical treatment." Denominational church? Renee had joined a Church of Christ, who have a long history (North 1994) of attempting to reunify all Christian denominations within one practice. Underscoring an issue that was implied in many of the student narratives, almost no one recognized themselves as part of an extremist movement. What this means from Renee's case is clear when we consider what such congregations have to say about evolution. Although Churches of Christ are "autonomous," the social norms they adhere to are not. With a literalist interpretation of the Bible, each is squarely anti-evolutionary. Four Creationists in my sample and one Both/and student were brought up in this denomination.

Herself a Creationist, Mary clued me into who has issues with evolution. "Probably really strict, religious individuals." Andrea also was clear on this. "Well, obviously religious groups. And I would also say just other scientists." As Andrea attended the same church as a prominent Creationist leader, she sees a disproportionate number of scientists at her church supportive of Creationism with a social support network to foster such marginal views toward science. "Obviously, they are researching deeper into the way things work and realizing how this is so completely complex—and how could this have ever occurred from nothing? So I think a lot of that is really impacting the scientific world right now." I now turn to two cases that exemplify the difficulty that such muddled perspectives cast upon educational practice.

7.3.4.1 Vignette: Litmus Test

As far as expectations go, I should have learned my lesson with Mrs. Weidman's example. My interview with Mrs. Garcia showed me quite clearly that when one goes looking under rocks, one often finds worms. I sat down with Mrs. Garcia during her planning period one day. Garcia had come to Salt Creek High School from Green High school in a neighboring urban district. Salt Creek and Green High were part of a different metropolitan area than Mason-Dixon State, over 100 miles away. Tanya, who had Mrs. Garcia for biology class at Green, clearly remembered evolution not being explicitly taught. As she recalled, when it was brought up by students it was relegated to backing material that was tersely explained and quickly packed away.

Garcia had taught biology at Green before moving on to Salt Creek where she now taught general science, for them a sort of ninth-grade introduction to physics and chemistry. There were early signs in our interview that the ideological charge of this teacher's compass was going to waver to a different position than those I had spoken with before. Talking about her experience growing up through Green High, I sensed this:

> I went to the Green High. It was a little bit of an alternative high school and people... *you know* controversy in the mainstream was not necessarily controversy there. So topics such as biological evolution were encouraged as classroom conversation, so I wouldn't say it was controversial there. We were allowed to discuss that. It was open. So if I remember correctly, we probably spent a month on that there. Now, in college it was much more controversial.

For other reasons, I was familiar with Green High, its clientele, and its sociological role amidst the other schools in this city's large consolidated urban district. By reputation, Green was where a certain subset of the school district's teachers tried to place their own kids, and was a popular spot for local university family children. It was a sort of private institution within a public system, for those with the cultural capital to negotiate the system. But hearing this school, which consistently performs in the top ten of all public schools in its state on standardized tests, referred to as "alternative" immediately disclosed to me an othering which I was not used to.

As Mrs. Garcia continued through her college experience, "People had very strong opinions and we discussed [evolution] and we learned about it, but I'm not so sure...." Garcia paused and reframed what she was saying. "I went to a Catholic University, so the opinions were just strong. Everyone was allowed to discuss their opinions and it was taught, but I don't think we spent nearly as long on it." As she continued through her education by discussing her graduate training, a perception of conflict is clear for her:

> We talked about it in graduate school. We talked about it in the science methods class a little bit and how teachers were worried about teaching it; and whether they should even teach it or should not teach it. The teacher started the conversation. I believe we spent the whole 90 minute class period discussing the importance of at least approaching the topics in class, and at least letting the students be taught the concept of evolution regardless of the teacher's personal beliefs.

Naturally, the discourse of this class was interesting to me so I asked her how people talked about it. What was the receptivity of the people who were in the class? Inadvertently during this, I uncovered the epistemological commitments of Mrs. Garcia. "Mixed. There were teachers who strongly believed in not ever teaching it. You know, it is not a confirmed theory. But yes, at least teach it and the student can make their own decision."

I steered the interview toward her teaching evolution. Having returned to Green High to teach biology as her first teaching position, this would now connect with the experience mentioned by Tanya. As Garcia detailed this experience:

> I think it went fine. I just kind of modeled it off of the way I remembered my mentor teaching it. I didn't do it very long because I wasn't comfortable teaching it, and I felt like I was okay leading a discussion and asking probing questions. But I didn't want the kids to control the classroom either. I decided that I was okay with teaching it, and at least leading class discussions on it because when you're teaching [evolution] it's much more of a class discussion concept than it is teaching it.

Garcia was losing me. What was this "it's much more of a class discussion" business? Was this due to wanting to detail the nature of science and as a best-practices approach, open up a dialectic? This became clear as Garcia more explicitly detailed what she covered in class:

> I don't remember the exact stuff we covered—Creationism, biological evolution, human evolution. I think there was one or two more things. If I remember correctly, when we started the conservation I brought it up, and I said, 'Now go home and discuss it with your parents and then come back and we'll discuss it tomorrow'. Whatever the kids brought from home in their home life and that they thought was true...we just went into it and I let them lead the class that way.

As she was on a roll, I very quietly encouraged this reflection to continue. She turned to the second part of a question I had asked, whether it was contentious in class:

> The biggest contention was that it just didn't exist or it wasn't—there's really no substantial evidence for [evolution]. I do remember one parent wrote a letter to the principal saying that they didn't even want their child in the class and they wanted an alternative assignment. And I do remember that they had to go to the library and they wrote a paper on something else. I forget what that was now. I did do an alternative assignment because [the class] cumulative assessment or some of the assessment was writing a paper discussing the different theories surrounding that.

As Garcia continued, the oddity of her detachment from the discussion continued with her choice of pedagogical strategy:

> It was much more student-led than anything else. After a week to two, I had them write a paper on their beliefs. I remember I let them research a lot of it online. We went to the library, and we let them research. I remember entering it and discussing it. There was one whole class for a discussion, and we reserved the library and let them go back and research what we discussed, and I let them research it some more, and then we wrote the paper. So it was over a week or so we just kind of kept bouncing back and forth. They would discuss, and then they would go "I'm gonna go look that up to prove you wrong." It was that kind of thing.

Stanley Fish (2005) wrote about the dangers of Creationists appropriating the language (and in this case, the pedagogical strategies) of the postmodern left, and in such form, this case was beginning to nauseate me. As if there could be no other possible punch line, Garcia unveiled the grand philosophical rationale by which evolution gets this special student-centered treatment in her class. When I asked her how she would teach evolution in the future:

> If I taught it again, I would teach it the same way I did. I would allow the lesson to be student-led because I'm a Christian. I believe that—you know—we came into existence in that way. I don't believe that we evolved in any way.

Smiling, we moved on to her current teaching appointment. Mrs. Garcia had taken a general science teaching position, the sort of integrated class content that many districts construct to be the foundation of later chemistry and physics classes. As I interpreted from the class notes on the board, the Big Bang was the topic of the day! Spellbound by how this topic might also prompt controversy, I inquired how she handled this topic. She began with student questions:

> The students were like...'well how do you know that a big bang created it' and I said, 'I don't, but I believe that somebody else did and he's the Almighty Lord' and that's okay. It's my flair as a teacher—when I'm teaching then it's my flair as a teacher to express my personal opinions because I'm a person. I teach it because the state says I need to, but I could also express my opinions. I'm going to test them on the Big Bang Theory, but I also can express other things.

Knowing that her current position did not require her to explicitly cover biological evolution, I was curious whether "her flair as a teacher" had bequeathed her insight as to how evolution was treated at Salt Creek High. Discussing whether evolution had been controversial in her years at Salt Creek:

> When I first started working here, in my interview they said 'how do you feel about teaching evolution'?, and I said 'well, if I have to teach it I have to teach it. I'm okay with that you know—I've taught it before', and they said, 'Well, we don't teach it here'. I just said, 'Okay, well then what's the point of asking then', you know, and then the principal just kind of laughed. She said, 'I just want to know your opinion on it'. And I said, 'Well, I just—I teach what I'm told to teach'. After that, once I got hired on I said, 'Hey, what's this all about'? And they told me the history.

What was "all this history about?" While interviewing, there are moments when time slows down greatly. Not wanting to disclose too much of my vicarious glee at having Garcia so blatantly unpack dirty laundry, I flatly enquired. "Really? What happened?"

> About six or seven years ago there was just a big community blowup about—I don't think there was a lawsuit filed or anything, but there were threats of lawsuits being filed that you can't make my student learn evolution. One or two classes were teaching it, and if you're a parent and you don't want your child to be taught something, you have a right to ask for alternatives. Whole classes were asking for alternative assignments. So just for the sake of saving face they just squashed it.

As she explained, local Southern Baptist and Pentecostalist churches had organized this protest. At this point, my intended line of questions regarding the extent and quality of evolution education at Salt Creek was moot. I redirected my questions into the general climate among her colleagues regarding evolution:

> Well like you know, in the teacher's lounge talking with my friends, yeah—I would say it's highly controversial. We have a very strong Christian network here at school. I'm a science teacher, and I'm also a strong Christian and I teach the concept of evolution…well, when I was a biology teacher I did—it doesn't mean I necessarily believe in it, but I believe in teaching.

The contradictions notwithstanding, I bravely enquired as to her thoughts on evolution's inclusion in the state's curricular content:

> If the State Core Content wants to have it in there it's their say so, and it's my job to teach it, and that's okay, and that's what I'm paid to do. I've had to teach other concepts that I don't necessarily believe in. It's okay to put my flair on it when I'm teaching like the Big Bang Theory today. And when I teach the Big Bang Theory you know—I'm questioned on why I teach that. Well, I have to teach the State Core Content. So it's a highly controversial conversation that we might have; 'why do you teach that…or why do you believe that and do you believe that. Well, what are your beliefs if you teach it'?

Having gotten plenty of this teacher's "flair" for evolution education, I moved on to what I expected to now be the easiest part of the interview—questions of worldview. As Garcia detailed she was raised Catholic, but along with her husband had changed religious practices. As with many Christian exclusivist students I encountered, the Foucaultian "denomination" game would once again set the terms of our dialogue. Once again laden with awkwardness, the transcript:

> I was raised Catholic
> *Okay.*
> But as of right now I'm Christian period.
> *Okay.*
> There is no denomination.

Tired of this synecdochic *coup d'eglise*, I forced a bit of history and context into the discussion:

> *Theologians and religious historians talk about denominations in terms of the history of Christianity since the Protestant Reformation, splitting with the Catholic Church and such. In my research, when people take an identity regarding religious practice, if they say they're 'no denomination', that 'nondenominationalness' itself becomes a kind of denomination. So is this a form of evangelical Christianity?*

I attend Northwest Christian. I wouldn't say—I don't think it is a denomination. It is serving others and spreading the word of Jesus. It's not a denomination, it's a new concept. It's not a denomination. It's Christian. It's a new concept that people don't understand. It's Christian. It's all inclusive of your believing in Jesus.

I was well aware of Northwest Christian Church and what being "just Christian" meant. As one of the nation's ten largest churches, part of the evangelical "mega-church" movement, it garners the pejorative label "Six Flags Over Jesus" by less-enthusiastic members of the local community. With a doctrine of biblical literalism, it also shares the distinction of being anti-evolution along with most of the top ten mega-churches. Just to confirm this, I asked Mrs. Garcia whether there was any conflict between the teachings of her religion and evolutionary theory. "Oh, it conflicts completely. Christianity and Catholicism do not believe in evolution period. It doesn't work. There's no such thing. God created the earth in 7 days and that's it."

With planning period time winding down, and being pretty damned sure that Mrs. Garcia might not make time to schedule a continuation of the interview, I got to the last question on my interview protocol. Did she see a relationship between a person's political affiliation and their attitude toward evolution?

Yes, if you are a Christian and of the Republican party I would say yes. No, if you're not a Christian. Like we have some Republicans who are Republican by association you know, and so they're like, 'Yeah, I'm a Republican but I believe in evolution'.

Although my own political hopes tended to see the USA moving toward a more Scandinavian model, my simultaneously libertarian tendencies were irritated. After first claiming "just Christian," now Garcia would claim the Republican cultural center from which *she* had "some Republicans" violating the purity of the core by "association." Smiling through what now felt increasingly like sharpened teeth, I asked:

Do you think a person needs to be a Christian to be a Republican?

No. No, I don't…(*with a tone of slight panic*) not at all. I'm just saying in every party you have every kind of person. That's what I'm trying to say. I'm a Christian Republican. I have friends who are non-Christian Republicans and that's okay. So that's a yes and no question. That's what I was trying to say. No, I don't believe you have to be a Christian to be a Republican. I just think that you know…it's hard to answer that.

Like the Christian exclusivist students before her, I once again found little hope from such interviews that there was any inkling of insight that a more pluralistic stance in society might be of our mutual benefit. In every other interview I held, I would close with an open ended catch-all question. This would give the interviewee a chance to share whether the interview prompted anything that they felt was important about the topics that we talked, but of which I had not specifically asked. Garcia was my first and only interview to quickly add to her narrative:

I just think that as a teacher it's important to introduce to students all kinds of concepts including evolution and that you should never not teach something because of your personal beliefs…and even though I don't believe in evolution it doesn't mean I would never not teach it.

7.3.4.2   Vignette: The Evils of the Secular Humanists

I met Mrs. Petit with trepidation. Petit had been Tyson's homeschool biology teacher, and as I knew something of his classwork, there was no doubt that this interview would detail Creationism. As a homeschooling cooperative teacher, she had conducted science lessons for a handful of homeschoolers that came to her house. There was no surprise regarding evolution. A member of a statewide network of evangelical homeschoolers, Petit came to the interview with reference textbooks in hand. During his education, Tyson detailed that not only had he covered evolution in his schooling, it was the major symbolic organizing principle by which his family and church community fended off the influences of "an evolutionary worldview." Given this knowledge, I carefully shaped the direction of my interview questions.

What was a bit of a surprise for me was, given the fairly nuanced understanding the Tyson held of biology as a college student, that his co-op teacher was more or less untrained in the natural sciences. As Mrs. Petit described it: "I do not have a mathematically inclined mind. It's a uphill battle for me to think mathematically. And science and math are so closely related that some of the sciences were a challenge for me." As she described it to me, although she enjoyed "fiddling around with the elements in chemistry," she preferred biology because "chemistry was math and that was hard." Having just stated what 1992's "Teen Talk Barbie" gained infamy for inscribing about gender roles, I was a bit nonplussed. I quickly regained my train of thought once I remembered the sociological world I was entering, albeit briefly.

Going on and graduating from college with a degree in early childhood development, Petit described her exposure to evolution in college:

> The professor, who had been married I don't know how many times, had said, 'Well, it's in our evolutionary makeup to have multiple mates, and so it's impossible for one man to have one spouse for life'. And I just remember thinking you are an idiot. You can't blame evolution on your lack of ability to stay with one person. So that—it was just that's part of the worldview.

As would sink in to me very clearly, worldview was not only an issue for Petit and her church family, it was a rallying point and *raison d'être*. Processing her professors sociobiological perspective on human sexuality: "I mean, it's that much engrained in our culture that it yeah...that it probably dates back to our 'Cro-Magnon man blah, blah, blah'...so I remember that coming up in college and thinking that was stupid."

A United Methodist until she was 33, the primacy of worldview appears to come into Mrs. Petit's life as she enters the practice of home schooling:

> I went to a seminar because as homeschoolers we go to seminars to learn how to use the curriculum that's out there and to learn. I sat in a seminar with a friend of mine and we were just starting to home school our kids—and this guy came up with a big black T-shirt and a snake and it said, 'Evolution is a lie'. We both went, 'It is'? I mean, we were just shocked that it was—he was so strongly against it, and we were kind of turned off. And then, we're like 'get over yourself, buddy'. And at that point, I realized, okay, I need to make a decision about what I believe and I need to know why I believe what I believe.

As she described, this re-evaluation of what she believed became an eye-opening experience that would change her perspective on everything in her life:

> Evolution was never this big controversy in my younger days that there was no big deal with it. And I have to admit I was not at that point in my life—I was a believing Christian, but not really reading God's word regularly. So just taking other people's opinions and saying, 'Oh, okay. That's what I should believe'? Not really going back to the source of what— I should be reading what God says about what I should believe, so.

Worldview became the foundational analytic of Mrs. Petit's life. "Everything is kind of tied back to your worldview." She went right for the big issues: "What do you believe about the origins of the universe? And what do you believe about the nature of man?" In an excited tone whose angle I rarely see close at hand in the discourse of my private life:

> Oh, [evolution's] everywhere. I didn't know I was encountering it, but it's everywhere, in every Disney movie, in every sitcom. There are references to the theory of evolution everywhere you listen and look, and when you—when you have your ears piqued to listen to them, *those* things. You pick up on them when you're aware of the subtlety sometimes. So I think it's everywhere, and I think it has been everywhere for a long time. I just wasn't aware of it.

For Mrs. Petit, evolution was an outward symptom of a fallen world being tinkered with and laden with the traps and snares of Satan. Moving to an evangelical church, she read vociferously. "Having been challenged to look at evolution as is it a lie or not, I began reading everything I could get my hands on, both from the home school community and from the secular community and reading the Bible." For her, the course she set was of the utmost concern as the fate of her children's souls lay in the balance. "I wanted to know what the truth was, and what to do. I had children."

With children nearing school age and partially at the request of her husband, she began the process of homeschooling that would find her as a co-op science teacher. "My husband has always wanted me to home school. He thinks I know our kids the best, that I'm a great teacher, he's got all of this confidence in me." Returning to the philosophical core, "one of the reasons that we home school is to be able to teach our children with a Biblical worldview." As she explained to me, easily 70–80% of homeschoolers she meets are members of evangelical churches and communities. Explaining this high percentage: "For the evangelical Christian community, I think homeschooling has exploded just for reasons of wanting to teach worldview."

Mrs. Petit came to actually teaching the co-op science class as a matter of practicality. Although informal, the network seems to put all the pieces together. With her daughter approaching high school age and anticipating college in the future, the issue of adequate science preparation became an issue. As her social network of homeschoolers discussed, "'Well, why don't you get a couple kids together and just—just see if anybody wants to do science together?' I am not a science person, but this book is written for homeschoolers to homeschoolers. And I learned along with those kids." As she continued with her anxiety:

> I was shaking in my boots because I thought I don't know what I'm doing, so we got some classes on video from Bob Jones University. I just didn't think I could teach it. And then, some very wise people said, 'This is the best homeschooling science book out there'. It's a whole series. It starts out in seventh grade and takes them through high school.

And so it was, and it was good. As Petit recalled, "I said well, Lord, if you will it, if you will give me the kids to teach, I will do the best I can with it." Recalling the Lord speaking to her:

> He honored that by giving us a great class and all of them were really good students. So many were smarter than me. I went by what Dr. Wile did. I took his hard work and we applied it. We read it. We studied it. We applied it. I chose carefully. I didn't just take a book off the shelf and say, 'This will do'. I chose carefully from someone who had credentials. He has lots of credentials in the scientific community and so didn't—didn't leave out evolution.

Jay Wile's et al. (2000) biology textbook, *Exploring Creation with Biology*, is a mainstay of the evangelical homeschooling movement. Filled with a repertoire of Creationist arguments dating back to Henry Morris's (1974) *Scientific Creationism*, Dr. Wile brings his expertise in nuclear chemistry to write a biology textbook. Worldview is a consistent theme throughout this book and others used for class. Again Petit focused on worldview regarding curricular materials:

> One of the foremost writers has been David Quine. He has a curriculum out called Starting Points and our kids have gone through, where you examine what worldview is. You examine the scripture for worldview and then you look at some books—like you read Frankenstein and you read Dr. Jekyll and Mr. Hyde and you examine them for worldview. But they're—they're challenged to come up with their worldview paper. That's what they do at the beginning of the class. Jay Wile also talks about worldview. Everybody—that's the buzzword because that's why we're doing what we do. We want our kids to go through life with their Biblical glasses on filtering everything through what we believe to be the truth of the Word of God. So it's all over the place.

As Mrs. Petit began to speak faster and with little break, her dialogue seemed building to a triumphant conclusion. Moving effortlessly from worldview to her philosophy of science, Petit hit another common trope of Creationist apologetics:

> True science is repeatable and observable. It is—and you cannot repeat and observe creation. So you have to go on what clues are left and it's a matter of faith to believe the Bible. It's also a matter of faith to believe that we went from primordial slime out onto the banks and formed lungs. And even the most staunch Evolutionist must say there's a point at which you must take a leap of faith because you know what? No one was there!

Common to almost every one of Ken Ham's apologetics lectures, "No one was there" is used to reduce the nature of evidence in science to the limits of one observer's immediate perceptual horizon. This homeschooling science teacher continued her rationale without a breath:

> And the more I learned about genetics, I realized if it's not in the genes it's not gonna happen, even if it's mutation. Mutations are bad things. They're not good things and they don't make a species stronger. You can't go from one kind to another kind. It just made sense to me.

As what I intended to be a short interview drew on, I sensed the dramatic finale was about to be delivered. I was not disappointed:

> I think much of science...the part that's not repeatable and predictable and that kind of thing, the stuff we have to make theories about—I think it is very important that you understand that it is a faith. It's a faith. It's a belief system, whether—if you're an—if you're an agnostic or if you're an atheist and say— Richard Dawkins, he has a belief system. He's not

an atheist. His God is evolution or his God is science. It's what he is serving, what he is lifting up, what he is—so I think when you're studying from an anthropological viewpoint (*pointing at me*), the science trends of the culture, you have to remember that at some level, it comes down to faith and what are you gonna choose to believe?

The small sample of teachers that I had spoken with surprised me by just how easily and deeply I found antievolution discourses to be prevalent in American classrooms. Half of them did what I considered to be an average to exemplary job, but the receptivity toward their work was similar to that of the classrooms at Mason-Dixon. Creationists in class would simply sit quietly and wait out this intrusion of evil. On the other hand, at least four other institutional settings disclosed strongly anti-evolution dialogue as the dominant discourse, with little science literacy regarding evolution built within them. At its worst, some classes and in some cases school administrations were perhaps breaking laws, ones surely unlikely to be enforced.

One surprising theme which was present in student discourse and also extended in teacher experience was the degree of antievolutionary sentiment within and about Roman Catholics. Although my interview data were not deep enough in these regards, it is likely that a study unto itself into this cultural construction might be well guided by Bloom's (1992) assessment of American religious practices. On the whole, the distinctively American faiths of Southern Baptists and the Latter Day Saints have had a profound effect far beyond the presumed boundaries of their social practice.

So as to show all is not lost, as a closing vignette, I did meet with one exemplary biology teacher at a girls' Catholic high school. Excited to meet with me, Mr. Joseph walked me through what was easily the most evolution-centric curriculum I had seen during my project, far surpassing that of Mason-Dixon. With evolution as the overarching principle, everything was on the table and clearly discussed. Human evolution was a matter for discussion from the first week of class. Demonstrating this, when I met with Mr. Joseph after school one day, he had a full laboratory of students working on an extra credit assignment in the adjoining lab. Gleefully walking me over to the door and cracking it open, he got their attention: "Ladies," everything fell silent as they looked up. "Nothing in biology makes sense…" his pausing and hand gestures cued a parroted response (*with a good amount of instructive eye-rolling at the stunt*) "except in the light of evolution." It was an old-timey technique, but against some of the other classes I had just seen, I once again had hope. For this one glimmering case, Dobzhansky would be proud.

7.4 La Pièce de Résistance

At the onset of this project, I delineated a rarely acknowledged detail about the social reception to evolution in particular, and about science in general. The factuality of science as a description of the natural world says nothing of people's interest in it. That evolution is coherently "real" does not by itself explain anything about social receptivity to the idea. Students are to varying degrees pulled between the cultural commitments of their religious practices and that which their science teachers and professors might have them believe and understand. Many quite simply have other interests which command their attention. The ends to which they serve have little to do with the nuance of theology or science. Faculty themselves are varyingly pulled between these epistemological domains, and as Chap. 8 will detail, teaching epistemology by itself as an explanatory framework adds to a conceptual fog in science education which might best be blown away.

Given our poor ability to articulate an understanding of each other, religious, scientific, and otherwise, one student case still stands out for the exemplary ways in which the rationalism of the "scientific objective," through education has failed. As such, we having investigated students, college faculty, and now teachers in practice. We now consider Laurel's story. At each turn, the means by which science education policy makers and science education theorists hoped evolution might sink in was lost on Laurel. Herself a Creationist of the mildest variety (she allowed for the evolution of nonhuman animals), she was preparing to become a middle school teacher of language arts and social studies. When asked of her interest in her biology class, she feigned just the slightest interest, as she expressed: "I really don't see what this will have to do with my future work." Her further rationale might be seen as practical or depressing depending on your perspective:

> Honestly, I put biology off to the very last minute because I was dreading taking it. And I had to take a bio, or a science with a lab, and a friend had taken it and said it was most likely the easiest from the selection of sciences we had.

Even when asked if the class might be a help in working collegially with her future science teacher colleagues, Laurel shrugged.

Laurel attended a public school in the general region of Mason-Dixon State. This school, like those in the region were one of a myriad scattering of balkanized micro-public school systems, ensconced within neighborhoods delineated by highly charged racial and socioeconomic relations. Regarding the quality of her prior evolution education: "Evolution wasn't taught in my high school biology class." But this exclusion did not negate evolution from Laurel's curriculum. "In high school, we learned about evolution... how we've been having a controversy of whether to teach creation over evolution." As Laurel detailed, this took place in the content of Mr. Woodman's earlier discussed world religions class.

After high school, Laurel attended two colleges on her way to Mason-Dixon. Of note, Laurel, on the advice of her mother, took a human evolution class. She was then the only student in my interview set who had had such precise curricular experience with such potentially contentious a topic. This course was at a university known for its physical anthropology program, having one of the field's seminal figures teaching on staff. Of her experience, Laurel still did not buy that human evolution took place. Laurel's professor at the time, then a graduate student and now a professor of paleoanthropology at a quite prestigious university, detailed to me in an interview a quite robust coverage of the field. In what was one of my oddest experiences in both interviewing and just plain trying to discern whether it was perhaps a case of *non compos mentis*, I pressed Laurel a bit more than other students. I asked her to detail to me her experience with the human evolution class. "In fact, it ended up being probably the hardest class I've taken in college so far. But it's probably one of the most enjoyable classes that I've taken so far. Yeah, it doesn't make sense at all, but...."Wanting to get her standpoint regarding evolution, I cut right to the chase. Laurel, *do you think humans have evolved*?

> Well, in a way I believe it, but in a way I don't because I—you know I'm Catholic—the Adam and Eve thing. And that doesn't really coincide with that belief, but maybe it does if it happened in a different timeframe. But there's no, nothing that proves when and where everything took place, I guess.

I would later return to this same point in a second interview with Laurel after her biology course had covered evolution. Reminding her of what we had discussed already, I was curious where she now stood. Had humans evolved along with animals and other life on earth, or had we been specially created? Laurel sided with the latter. As we talked, it became clear that the narrative and importance of religion was simply more resonant in Laurel's life. Having the common "only a theory" conception of theory as used by many in confusion regarding science, Laurel was no different than a large majority of students I spoke with and in that same way similar to most American adults. As Laurel charged about how religious knowledge trumped evolution:

> I don't know about my dad, but my mom definitely, I think, feels the same way that I do. The majority of how evolution and our religious beliefs could be somehow intertwined. But as far as evolution goes to most of my other family members, they'd probably deny evolution all together because they're hardcore Catholic.

Intertwined yes, in that some evolution had taken place. Intertwined no, as for now for Laurel, human beings had not evolved.

7.5   Evolution Is Political

Although a thematic undertone throughout the entirety of this project, it bears a clear mention—if only in short. Evolution, at least as the concept is deployed as a piece of conceptual capital, is political; it is a strong proxy marker for predicting political identity. This is not likely surprising or new information, but the clarity in which certain ideological and political identities lined up in this project was uncanny. In Kentucky, where I spent a good number of years, it is common knowledge that all Bourbon is whiskey, but not all whiskey is Bourbon. By analogy, barring one discrepancy, all Creationists (at least in my data set) are Republicans, but not all Republicans are Creationists. Seen below (Table 7.3), to be a religious exclusivist was to be a Creationist. Every Creationist I spoke to was an active member of a Christian exclusivist congregation.

What this type of data portends has much to do with the future of anti-evolutionary trends in the USA. In an early 2007 Republican presidential primary debate, three

Table 7.3 Evolution and political affiliation

		Exclusivist	Both/and	Agnostic/atheist
Students	Republican	8	5	1
	Democrat	1	8	1
	Independent/other		3	4
College faculty	Republican	1	1	
	Democrat		5	2
	Independent/other			
Teachers	Republican	3	1	
	Democrat		3	
	Independent/other		1	

of ten candidates raised their hands when asked if any of the candidates did not "believe in evolution." Candidates quickly glanced around to see who had voted and who did not, with constituents watching. Each of the no votes came from candidates who either attend or are members of Christian exclusivist congregations: former Arkansas Governor Mike Huckabee—Southern Baptist; Representative Tom Tancredo—former Roman Catholic and now evangelical Presbyterian; and Senator Sam Brownback—a converted Roman Catholic, but who attends a non-denominational evangelical Christian church.

Mooney's (2005) thesis summarizes the political impact of the Republican Party's relationship toward science. He sees it as dominated by many who either ignore or do not understand science, and by a smaller set who willfully ignore or downplay science that may be damaging to short-term political interests. For those interested in increasing the rate of Americans who have no quibble with evolution in classrooms, some recent trends in American religious practices may warrant further investigation. Although anti-evolutionary attitudes may have once been typified as something coming from small rural churches, there has been a stunning shift in American religious practices toward evangelical Christianity. As of 2007, nearly 30% of Americans counted themselves as members of an evangelical congregation, with one in ten Americans attending such churches every Sunday (Pew Foundation 2008). Although there is by comparison a tiny evangelical political and theological left, on the whole, this is a politically and theologically conservative movement. These shifts are an area of future interest for researchers interested in where anti-evolution attitudes are being fostered. The current growth of evangelical mega-churches is typified, in my check of the largest ten's doctrines, by anti-evolutionary attitudes. Evangelical mega-churches, like American middle class residential patterns, have tended to spring up in new-growth suburbs of mid-sized and large metropolitan areas.

Leaving Mason-Dixon State and the students, faculty, and community members with which I spoke, we now turn toward synthesis. How, given the value rationality and differing ontological positions from which people often avoided or deflated evolution, can we begin to look at evolution education with fresh eyes and toward a useful dialectical education?

Chapter 8
Darwin's Hammer and John Henry's Hammer

When John Henry was a little baby boy
sitting on the his papa's knee
Well he picked up a hammer and little piece of steel
Said Hammer's gonna be the death of me, Lord, Lord
Hammer's gonna be the death of me

John Henry (excerpt)
American Traditional

I set out to ethnographically answer this question: how does one learn about evolution? Past the interpretative answer I have presented, the narrative can also be followed as a metaphorical linear journey. Not just a simple matter of temporal succession, but as I would have you think about it, a sort of rail journey toward an end which I will disclose. Underneath us are two rails that I have laid down: The rail of religious *Truth* as clung to by Creationists, and the rail of scientific *Truth* as clung to by a certain conception of science. Both are needed to define the route and to keep the train from derailing. These rails are entrenched in the American cultural experience. Epigraphs posted along the way from traditional American song gave us our soundtrack. Stories about trains, journeys, salvation, redemption, and damnation, along with a few from outside the USA; I have left these markers along the way as concrete reminders of our cultural heritage in music.

As your engineer, I have driven deep into the mythos of American hope in both religion and science, faster and more furious amidst fire and brimstone smoke. We will come to the end of our journey at a scene from the heart of the American relationship between science and religion enshrined in our folk tradition. Where once John Henry drove steel, Creationists will take a stand. Where once a steam

D.E. Long, *Evolution and Religion in American Education: An Ethnography*,
Cultural Studies of Science Education 4, DOI 10.1007/978-94-007-1808-1_8,
© Springer Science+Business Media B.V. 2011

drill was, now will be the hammer of evolution. As all railways eventually come to an end, we will finally leave our mythic journey, stepping off our train. Pushing forward into new possibilities, we will consider a view from another world. Experiencing the disorientation of this space, we will have to come back to this world.

Moving out of my ethnographic description of one slice through American educational practices toward evolution, we continue with this chapter to move up to my final level of analysis and come to an end. I first detail how existential phenomenology can more fully represent the experience of Creationists as they reject evolutionary theory, with particular attention paid to the practical rationales within which these views originate. I next illustrate these rationales exist within social fields, and how ontological positions limit one from coming to acquire some knowledge through education without significant social consequences. Using the work of Heidegger (1962 [1927]), Bourdieu (1977, 1998), and connecting back to my earlier use of Swidler (1986, 2001), I work out a more robust cultural logic by which Creationists take an epistemological stand in a world, and what this implies for educational possibility in general. Lastly, I consider what it means to have your world of cultural significance erased from you, and how we might work through educational strategy to soften this transition.

8.1 The World Turned Upside Down

As a child, I always had a fondness for maps. With what I have come to understand as a strong ability for visual recollection, a sort of aptitude to memorize the visual arrangement of things in space—maps oriented my imagination toward what might be out there in the world. Rather early on, when spinning a globe in my bedroom, I remember being struck by suddenly unknowing what made the "up" on the globe "up." Greenland and Iceland were up, while Australia and Antarctica were down. There was a historical and etymological story which I well understood, but this momentary perceptive move was something different. What was this "upness," this orientation that was quite different than the everyday "upness" by which we temporarily hold our bodies against the force of gravity?

This perceptual shift, where up is not always up, and the basis of upness itself is an object of question, is the kind of area of inquiry for which Edmund Husserl began the work of phenomenology. For Husserl, the "naturalistic attitude" of the mathematically described sciences as he saw it would need to be suspended, "bracketed off" from the world of human perception (Husserl and Welton 1999). This was required to apprehend a discrimination of the real from one's perception of the real. Working past the point-to-point limits of Cartesian space and the difference between the mathematical approximations of ideal objects and their phenomenal reality, Husserl began the project of phenomenology. The goal was to describe a science of our experiencing the world, for which the data of natural science is based in but cannot itself fully explain, due to its mathematically dependent ideation.

Taken in new directions by his most influential student Martin Heidegger, and later Maurice Merleau-Ponty among others, phenomenology exists today as a philosophical program of our experience, rather than that of attempting to master a final reality. As equally important for Heidegger as it is for Husserl, the project is not against science, but simply works to articulate that which we experience as reality from that which is atomistically "real." In the spirit of Hume, although we can describe the world, this says nothing about a way that things *ought* to be. But the ways that culture forms and directs narratives about reality *are* distinctly in play when analyzing receptivity toward evolution, especially those that claim access to one final description of it. As such, a handful of concepts from Heidegger's radical study of ontology underline my analysis.

This becomes important as we consider the basis by which Creationists reject or avoid evolutionary theory. If Creationist students are simply "misconceived" about the factualness of evolutionary theory, then showing more and more incontrovertible evidence would be the answer to Creationism *as a problem*. But as Chaps. 3 and 4 articulated, student commitment toward Creationism is not a matter of having a problem; such students have an ontology *exclusive* of evolution as a possible explanatory mechanism. *Being* a Creationist is a matter of identity, with its own set of *a priori* epistemological commitments. As there are no naturalistic means by which we can measure how far such a student is from a correct or "rational" reconceptualization, we need to step back and reconsider our intellectual method. The existential terminology of phenomenology offers ways for us to consider just "how far" this distance might be. As I submit, this ephemeral distance can be described using Heidegger (1962 [1927]) conception of "distantiality." Distantiality describes the concern we have as beings toward one another, in our case illustrated by students of differing ontologies encountering evolution in the sociality of the classroom:

> The care about this distance between them is disturbing to being-with-one-another, though this distance is the one that is hidden from it. If we may express this existentially, such being-with-one-another has the character of distantiality. The more inconspicuous this kind of being is to everyday dasein itself, all the more stubbornly and primordially does it work itself out (p. 164).

Classrooms are places where differing ideologies meet and are at times found to be in competition or even collision. When, where, and why one stands up for their position certainly depends on a great unstated gestalt calculus of social costs. In the university class, evolution might be afforded the dominant didactic voice, whereas for many Creationist students, their churches and peers enforce quite differing narratives. Sensing the narrative of home juxtaposed jarringly against the narrative of science, the dissonance produced can manifest itself as a form of anxiety, or more commonly as anger expressed in rhetoric. As Dreyfus (1991) extends Heidegger:

> Heidegger...link[s] this uneasiness with our deviation from norms to anxiety, and will interpret our eagerness to conform as a flight from our unsettledness—an attempt to get ourselves and everyone else to believe, or better, to act as if, there is a right way of doing each thing (p. 153).

If what we then seek is a norm, what would we expect to find regarding the teaching and reception of evolution in our classrooms? Returning to Dreyfus (1991):

> Norms and the averageness they sustain perform a crucial function. Without them the referential whole could not exist. In the West, *one* eats with a knife and fork; in the Far East *one* eats with chopsticks. The important thing is that in each culture there are equipmental norms and thus average ways to do things There *must be*, for without such averageness, there could be no equipmental whole (p. 153).

At least for the field of orthodox scientific practice, one proceeds within a referential whole in which evolution has taken place. This whole, as Heidegger would see it, is the referential totality in which we are all enmeshed. Like the fish that cannot see the water within which it swims, we are in worlds of cultural equipment with significance. "Norms define the in-order-tos that define the being of equipment, and also the for-the-sake-of-whichs that give equipment its significance" (Dreyfus 1991, p. 153).

What is clear regarding the social practices we have seen regarding evolution is that there appears to be not only conceptual differences between parts of Creationist and orthodox scientists' referential totalities. Orthodox scientific practice and Creationism serve ends of incommensurable ontologies. In this way, this argument has structural similarities to Kuhn (1970) and Feyerabend's (1987) discussions of the incommensurability of scientific paradigms. Moving between the two (Creationism and orthodox science), as we have seen in my interview prompting, cues one to experience the existential anxiety of stepping off into an abyss. The cultural "distantiality" of these ontological differences might as well be a billion miles in Cartesian terms. But before we go any further, we must get a clearer sense of what Heidegger means by referential totality, in-order-to, and for-the-sake-of-which. With this, we will begin to see more clearly the significance and rationale of avoiding evolution for Creationists and how this translates into matters of scientific understanding and identity.

8.2   Evolution as Equipment of a Scientific World

8.2.1   Darwin's Hammer

For Heidegger, individuals act in their social practices as beings in a referential totality, skillfully coping with the equipment of their worlds. Heidegger refers to us in this way as *dasein*, or "beings that take a stand on their being." To get a sense of what he means regarding the equipment of a culture, and how that gives us significance, Heidegger's (1927 [1962]) prototypical example from *Being and Time* is the hammer. The hammer, for a *dasein*, exists in a referential totality toward which hammers are for hammering. The hammer, as equipment of *dasein*, is utilized by *dasein*, toward a practice which *dasein* engages, in the easiest sense, hammering a nail to a board. The practices and equipment themselves that *dasein* engages in become "transparent" as Dreyfus (1991) describes it. We "skillfully cope" with all the stuff around us in life, almost never considering the thing as itself. Demonstrating this, the masterful *dasein*, a carpenter or craftsman as Heidegger saw it, goes about the practices of hammering, all the while having no conscious or "cognitive" sense of the hammer as present. As he or she might do this, the carpenter chitchats with a buddy, talks about a person walking down the street, thinks about lunch, masterfully hammering the nails and wood toward a building project for which one hammers. At no time, as *dasein* hammers, is there an awareness or cognitive equation for the mass and orientation of the hammer, its velocity, or does appropriate bodily comportment flash through the mind unless there is a breakdown. The atomistic description of the hammer, the amount of iron and carbon compound molecules that make it up, both is never considered and says nothing about the significance of the hammer as equipment in a referential totality of human significance. In short, what we are after is the point that science has no way to talk about the relation between the hammer and the nail, and the deep culturally significant continuum of relations that ripple out from them. In this way, this ontology of equipment is similar to Geertz's (1973) "webs of significance."

If equipment is not available, in this case the orthodox scientific conception of evolution for Creationists, Heidegger describes this as equipment being "conspicuous" (p. 104). Conspicuousness emerges in "breakdown cases" as Dreyfus describes, illustrated as when equipment either is not part of your world, or it suddenly shows up in the scientific sense rather than the cultural sense (p. 71). In this way, the hammer also becomes conspicuous if he or she misses a nail and hits their finger. As Dreyfus would describe it, there is equipment, that for some worlds, is conspicuous in that it is too heavy. Like a hammer made of "butter or pillows," evolution is essentially useless within the referential totality of the Creationist. What purpose for-the-sake-of-which might evolution serve a Creationist, for *being* a Creationist? This is not an issue of ultimate ends, more an ontological description of the styles of our lives. As Blattner (2006) describes of this view:

> I am a father, a husband, a son, a neighbor, a youth baseball coach, and so on. Are all of these for-the-sake-of-which projects that refer back to one ultimate for-the-sake-of-which in my life? There is a temptation to think so, especially when we are forced to deliberate about the overall shape of our lives...It is this sort of reasoning that led Thomas Aquinas e.g., to posit a last or ultimate end or goal of human life, a final anchor to which a final business of rational deliberation could be tied, when needed (p. 60).

Considering my earlier preposterous example (the mother, Creationist, doctor, atheist, and astronaut), such illustrations are jarring in what they imply about the temporal and ontological. One needs time to master practices in life, and some practices are rooted in ontologies which exclude the possibilities of others. Ideology then can be seen in some ways to significantly mediate perception. As such, for Creationists, evolution has no significance except portending doom. Evidence such as evolution, if taken seriously, potentially undermines a way of religious commitment.

As arrogant as it seems, the fact of the matter is that evolution is not readily available as transparent equipment for everyone. Creationists are not going to suddenly pick up Darwin's hammer and start building the structures of a more and more refined evolutionary synthesis. Dreyfus calls this ready-to-handness that Heidegger called *zuhanden*, "skillfully coping" with the equipment of our worlds. Ready-to-handness for Heidegger is the ability to naturally use equipment of a world within which it is naturalized, in that one would never question its utility toward a practice. But one must learn the utility of equipment. For one to acquire and then skillfully cope with evolution (in our case) as ready-to-hand equipment, or so that it becomes "transparent" as Dreyfus elaborates (1991) regarding equipment generally, requires a positionality toward theism *and* science open to multiple valences of philosophy. As I have discussed at length, the sociology of American religious life does not produce an even field of such situated people.

Individuals without such a positional stance, those unabsorbed into a vocabulary of understanding congruent with evolution, should naturally find evolution troublesome if not simply false. Creationists then act rationally *respective* of the referential totality within which they exist. Turned around, why would a Creationist—while *being* a Creationist—ever come to find evolution as useful equipment? Your or my discomfort with such resistance, if we have it, comes from our ethnocentric othering of this epistemic variety in American culture. Or turned back around to a more conventional way of looking at it, the ontological variety of American cultural life does not produce individuals equally suited to engage in a rational consideration of evolution, as a natural consequence of reasoned argument supported by scientific data.

Past these difficult problems, we then remain tasked with understanding the mechanics of those who do change, and what this means for science education, worldview, and identity. Rather than an end as some fixed gilded city on a hill, Heidegger saw our being as disclosing styles of life, which in our everyday comportment though life and our being with each other further discloses significance. These styles, our being for-the-sake-of-which, describe our ontology. As Wrathall (2006) summarizes of this view of a world of significance:

> Understanding inhabits a domain of possibilities. When I understand the world, I find myself in a particular situation, where I have available to me different 'for-the-sake-of-whichs', different ways to give order and purpose to my life. And when I settle on one of these 'potentialities-for-being', I will find that the world is set up to permit particular activities and objects to be used in pursuing this way of life (p. 28).

One can *be* a Creationist, *be* a teacher, *be* a student—but at some point the distantial space between some styles makes simultaneous ways of being too dissonant. As discussed earlier, one cannot simultaneously be a Creationist, mother, firefighter, atheist, and an astronaut. Some ways of being obliterate other possibilities.

How does evolution become part of one's referential totality? If we then think of our own referential totality represented by vocabularies distinctive of people populating social fields, we would then have utmost concern with the types of dialogue and rhetoric spun onto evolution in social practice, namely, through education. Schooling as a field of social practice becomes suspect in these regards, as critical pedagogy has shown. As Kincheloe (2008) has recently reminded us, the instrumental reductionism so common in today's day-to-day educational norms of testing and measurement are the tools by which epistemological complexity is often subverted. Evolution, as potentially controversial, may be easily enough shooed aside as seen in many cases in this book.

Perhaps more than any other, it is when Bourdieu adapts (Bourdieu 1977) Heidegger's vocabulary of ontological equipment to the social science field that the critique of class distinctions via education become so powerfully articulated. The "hammers" become the "capital" by which individuals, differently equipped as products of culture, emerge into the competitive arena of social fields. Bourdieu's anthropological philosophy of practice then lays out the terms by which class distinctiveness makes the capital of the dominant classes natural or inevitable, that which, of course "one" might usually strive to acquire and master. Therein lies the rub for Creationists. One may wish to gain the capital of "scientist," but one must carefully hide their secret Creationist identity, if they have it, if attempting to acquire the capital of a university degree in most settings. Tyson very carefully holds back his critique of evolution from his college professors lest they judge him. Renee clearly felt stung by her attempt to introduce Creationist interpretation to her peers. But as James's move from Pentecostalism showed, many often lose this commitment along the way. What is never discussed in an organized fashion is that those who find their religious exclusivist ontology moved do not, by and large, become religious inclusivists, they tend to become agnostics or atheists.

To what degree is the concept of evolution, as deployed in everyday education, Darwin's hammer? How do people, extending Bourdieu's market metaphors, "buy" it? As Bourdieu and Passeron (1977) show and Lareau (2003) recently develops further, one's ease with the language types, practices, and knowledge sets one encounters in the arena of formal schooling depends in great part on social class, race, gender, and cultural context—the ready-to-hand equipment of the learned and lived life. Where and with whom you attend a school, and what vocabularies and social practices you master matters. Geoffrey Canada's relative success in erasing achievement gaps of low-income children in the Harlem Children's zone is testament to this philosophy in practice (Tough 2008). But let us not slide back into seeing such "success" as a path toward a greater truth. Canada's program is simply bodily manifestation of an experiment in holistic lifelong socialization. Where we may all agree that it might be a socially just way to go, there is nothing final, more truthful, or transcendentally better about such educational ends.

Just as Willis's (1978) working-class lads learned to resist trading away their own forms of capital, Creationists resist investing in new forms of capital as they have no significant traction within the style of religious life they hold to utmost. In essence, this was one of Bourdieu's most useful points. The "misrecognition" of elite capital as somehow "better" than that of the cultural proletariat keeps a social hierarchy in order. The success in this case lies only in our seeing children from poverty act more like upper middle class children. Given all we have seen in this project, religious ontology also certainly works this way. What part does the informal home education of religious ideology play then, parallel to the above insights regarding the public's receptivity toward evolution? Given the deeply scored lines that religious practice and identity forms in many people, one might expect a great deal.

When we consider our earlier discussion and analysis of Swidler's strategies of action, and how the equipment of our cultural lives structures our possibilities, the overall picture becomes a bit claustrophobic in that we may come to realize ways our lives are actually quite socially limited. As Swidler (2001) summarizes nicely:

> Think of [someone] as a hiker ascending a mountain, with culture as a description of the path she follows. The mountain's topography will certainly affect her route. She will pay attention to a boulder she must cross or go around, to steep or flat places, to openings in the trees. But other features on the mountain that do not directly affect her climb may be irrelevant. She may misconstrue the larger shape of the mountain, yet well describe her own path (p. 132).

Thinking about a synthesis of the Creationists I spoke with, their social networks and circles of discussion were highly church based, and centered on insular practices that in many ways ironically excluded those that they might wish to evangelize. Neither Tyson and Andrea nor Cindy spoke of secular campus activities—the social connections they spoke of were dominated by family and church youth groups. Andrea, who spoke of having never really thought about other religious views, still had internalized a sense of opposition from the "world." Renee, who had become a "new Christian," already had expressed some alienation that her changing circle of practice had brought. Like Swidler's hiking metaphor, these students' paths were clearly articulated to them, although they grossly misrepresented the mountain. Enjoin with this our broader discussion using Wuthnow's (2005) ontological categories of religious practice which I used to animate my students, and one might begin to see clearer structural limitations in the ways that we talk to one another in cultural life, let alone the difficulty of discussing the contention of evolution.

The central challenge my line of argument raises is one of the stickiest kind for American science education policy writers. If public education curriculum writers are to design and assess a curriculum that contains evolution and requires that students deeply understand it, what kind of understanding exactly are we looking for? In many cases, as student experiences detailed, the hard questions that an understanding of evolution prompts for religious exclusivists makes a central (at least on paper) barrier of US education, the separation of church and state, potentially untenable. If comprehending the intellectual parameters that make evolution possible (geological time, mutability of species, and abiogenesis) require one to dismiss or seriously call into question one's commitment to a literalistic conception of Biblical *Truth*, then what of respect for religious freedom in practice? Can a religious exclusivist both be a Creationist in their intellectual convictions and be led through a state-mandated education that, via apprehending an orthodox understanding of contemporary science, alters their religious being through the process of public education? This line of questioning rarely rises from most scientists or the intellectual and theological left, as religious exclusivism is for them abhorrent, often seen as symptomatic of a weak mind, and certainly nothing that one could imagine themselves being. Some such as Smith and Siegel (2004) have advocated that science classrooms are best left bereft of any discussion of belief; rather leave that to a philosophy or religious studies class. While reasonable on paper, such a view simply does not honestly account for the fact that a religious exclusivist's ontology comes to class with them. The act of leaving belief behind, suspending the defining commitment of an exclusivist's identity, simply does not happen without ontological consequences.

Many such as Miller (1999) have posited that to be educated in concepts such as evolution has no effect on faith. Miller gently sidesteps or does not take seriously the importance of ontology. For religious exclusivists, to move to an ontological position where evolutionary theory is tenable, is to have moved to a position where science and faith are either non-overlapping magisteria (Gould 1999), or simply Biblical text now represents metaphor. In either case as argued here, both require, as a matter of education at the hands of the state, a religious exclusivist to change their religious practice. The difficult implication this line of argument poses for educational practice simply cannot be easily avoided, and simultaneously hold to an expectation that evolution education will somehow magically get better. Parents, students, teachers, administrators, and college faculty each have ontological positions regarding faith, most often not explicitly talked about. To ask religious exclusivists to accept or to design educational scenarios, where they come to accept that evolution has taken place is to put in motion educational policy recommendations that, when applied, passively advocate a person changing their religious commitments. The ugly and difficult question science educationalists will eventually need to ask is: can a public curriculum respective of the US constitutional establishment clause both get evolution education done and not tax religious exclusivists of their defining commitments?

8.3 "John Henry Laid His Hammer Down and Died"

After a long journey through one perspective on the American relationship between science and religion, we come to the ends of our tracks. But like all traditional Western narratives, we must have our conflict resolved. Having been supported by a rail of scientific *Truth* and a rail of religious exclusivist *Truth*, we now step down off our hell-bound train, as we have run out of track. We have run up to a wall of temporality, a "now" for which we must consider what we will do next; we must consider what future possibilities we will press into.

All along the way, these two rails have grounded our journey. All throughout this writing, you have seen the ways in which our cultural practices of schooling, devotion, and discourse conspire to level down what could be. For the most part, ministries such as Answers in Genesis and the Creationist students I spoke with were not against science *per se*; in their own understanding, they thought that a true science only could be possible while kept under the authority of the Bible. In fact, some Creationist students were genuinely excited about the instrumentalism of science. Likewise, culturally blind scientism seemingly cannot imagine a means or rationale by which such people could, or should even exist.

At its most general level, the American mythic experience is one run right through with the dual commitment of a hope in science and a hope for salvation. Like the story of John Henry, each is committed to incommensurable but eerily similar ends, each willing to risk all as if acquiring *Truth* was a race. A few steps down the tracks we can imagine the epic struggle coming into focus. The railway construction captain and John Henry, the steam drill or Darwin's hammer, John Henry or the Creationist—the battle in our case is the same. Facing the removal of his purpose, the erasure of what it means to *be* a steel driver, John Henry picks up his hammer. Swinging with all his might and filled with the spiteful purpose at having been cornered into a challenge, John Henry continues while facing the threat of the steam drill:

> John Henry told his captain
> Lord a man ain't nothing but a man
> But before I'd let your steam drill beat me down
> I'd die with a hammer in my hand

John Henry (excerpt)
American traditional

Like all good fights, and even more so given the attraction of cliché for the crowd, the current battlers over evolution and Creationism line up. The Creation Museum, Dawkin's *God Delusion*, Ben Stein's *Expelled* film, or Bill Maher's film *Religious*, each serves the same simple polemic of one rail or another. With the surety that their chosen path is the noble one, that the ends toward which their lives are set are those of the singularly divine, Creationists trundle on. Ever surer that the scientific foundation of Darwin's hammer is the product of the Devil's workshop, Creationism simply does not abate. But, as we know, the steam drill did go on to replace the steel driver of the early American railroading era. What it meant to *be* a steel driver had been erased from possibility. Creationism one day too shall pass, as all cultural traditions and their religions burn brightly and then fade:

> John Henry hammered in the mountains
> His hammer was striking fire
> But he worked so hard, it broke his poor heart
> And he laid down his hammer and he died

John Henry (excerpt)
American traditional

As I see this cultural succession, human existence is hallmarked by the turnover of worlds. Ways of *being-in-the-world* have equipment suited to the practices which make a world intelligible, but also, in quite an evolutionary way, open up the possibility of new practices in ways that education can be instrumental. As Dreyfus (2005) would agree:

> The new world with its new possibilities arises from the collapse of the old world and some day it too will die. That is, it will make sense no longer, become impossible, unthinkable and so give place to new forms of intelligibility (p. xxviii).

This collapse of all intelligibly, which I discussed in Chap. 3 in the work of Jonathan Lear at the end of the Crow Nation, illustrates the same. This collapse of a world, or simply contemplating the prospect of it as my interviews showed, induces a form of existential anxiety. With no other option than an absolute epistemology, if taken away, a hole in Creationist being opens before them with nothing but an abyss staring back.

Imagining the smoke clearing from the mythic battle, at the ends of the tracks, amidst all of the rubble, tools, and equipment for which we might consider our next move, we would be wise to imagine something else. Buried right there below the railbed would be what appeared to be—a third rail. Of the same common imperfect material, yet curiously covered up, I brush away to find not only a rail, but a warning sign.

* Danger * Third Rail. Risk of Death by Electrocution.

Knowing full well what I was doing, if only to illustrate a point, I go ahead and grab hold.

> For storms will rage and oceans roar
> when Gabriel stands on sea and shore
> and as he blows his wondrous horn
> old worlds die and new be born

Mother Shipton's Prophecies
Shipton and Mee, A. (1989).

"…We may well discover, in moments of despair, that we can no longer conceal our underlying anxiety over the seeming meaninglessness, banality and emptiness of devoting one's life to merely doing what one does in one's culture."

Mark Wrathall (2006, p. 61)

8.4   The Death of Prescribed Worlds

What does it mean to have all worldly significance stripped from you, for your very way of *being-in-the-world* to be shown as pointless, a farce? This deworlding of the world, to show one's way of being to have been superseded, "outstripped" as Heidegger termed it, prompts a form of anxiety at having had one's intelligibility left behind. This collapse of all possibility for a way of being is a form of death, a loss of a way of *being-in-the-world*.

Writing about the Crow nation, Lear (2006) makes a clear case of what was at stake when the end of a world came to pass. Lear, quoting Crow Chief Plenty-coups from an oral history with Linderman (Plenty-Coups and Linderman 2003):

'I have not told you what happened when I was young', he said when urged to go on. 'I can think back and tell you much more of war and horse-stealing. But when the buffalo went away the hearts of my people fell to the ground, and they could not lift them up again. After wthis nothing happened. There was little singing anywhere. Besides', he added sorrowfully, 'you know that part of my life as well as I do. You saw what happened to us when the buffalo went way' (p. 2).

The basis by which one might *be* Crow, the possibility of Crow *being-in-the-world*, had passed. As Lear details, Plenty-coups, along with many other Crow, went on to successful farming lives, but for the basis of this style of success, "you [knew] that part of my life anyway." Such cases, if less extreme, are not unusual in more recent times. As Blattner (2006) describes a view perhaps a step more familiar:

> The self-understanding of being a steelworker came crashing to an end in the Monongehela River Valley outside Pittsburgh, Penn. during the early 1980s. Someone who had understood himself that way no longer could; the context stripped it of its livability (p. 150).

Lear's analysis of ethics at the edge of cultural devastation prompts interesting questions for our case at hand. In the face of evolution as coherently true, as a paradigm of scientific orthodoxy, what should a Creationist do? What next steps respective of *being* a Creationist should one take? Lear terms this line of questioning "abysmal reasoning." What is the appropriate thing to do when one's strategic repertoire and entire grounding in the world begins to dissolve? As we work back from the Crow example, through the Creationist, to the more seemingly practicable example above from Blattner, the implications for education become clearer. Education for what? Education to be what? For whom? If we remain committed toward a view of science as disclosing *Truth*, we are left with the odd proposition—is it wrong to be a Creationist? If it is, where then is the line by which religious belief is right? Science and religion are so easily muddled together; many simply do not make clear separating distinctions in their day-to-day lives. As I have pointed out, ontology rather than epistemology is the appropriate level of analysis when considering the basis of Creationist complaint.

Returning to the rationale of those with whom I spoke, a Both/and student or faculty member might have shaken their heads in confusion, or just as likely feigned disgust, at considering the cultural logics of a Creationist. For the Both/and, to think *that* way is inconceivable. As Swidler (1986) would say of the Both/and position, ideology "had gone underground." The normativity with which the Both/and look unto the world is to cover up and have great faith in the stability of their inclusivist ontology. Pushing back against them, the Creationist looks outward toward the world and sees one great *Truth*. Everything cast in such terms leaves little room for matters of interpretation or accommodation. As I worked to edge Creationists at least toward a glimpse of other possible worlds, the *abgrund* or abyss of meaninglessness this prompted was too much. It scared, unsettled, disappointed, and, in extreme, prompted students to associate evolution with death. This groundlessness, the fear of loss of metaphysical meaning is clear for the Creationist, but no less unclear when one considers the broader consideration of why, apart from tradition and contingency, one "believes" anything.

My intrusion into the hearts, minds, and religious knowledge of a small set of Americans left me a bit shaken. Short of its explicit inclusion in public school curricula, most people simply do not have a means by which a dexterous knowledge of religious diversity might be introduced to them. Why would they? Unlike the socially restrictive practices of religious exclusivists, Wuthnow (2005) saw that those who had been socialized to have an ecumenically tolerant view of others were highly likely to have been cultural products of middle and upper middle class households, with much higher levels of education:

> Going to college, travelling abroad, and finding other ways in which to experiment with new lifestyles take financial resources, and thus it is not surprising that [Both/and] tend to come from middle and upper middle-class backgrounds rather than from working class families (p. 116).

Both/and socialization included concrete examples and discourses that worked toward ecumenical views of society.

Church people who are open minded toward other religions have typically come to this position as a result of social experiences that not only exposed them to people of other faiths, but also gave them a way of talking about these faiths in relation to their own commitments to Christianity. For many people, these ways of talking were learned as children. Their parents talked favorable about diversity and encouraged them to take a positive attitude toward other religions, the same as they did toward other races, nationalities, and ethnic groups (p. 135).

Such socialization explains Both/and ecumenical tolerance. But of the Both/and students I met, along with most of the religious exclusivists, I was surprised at their limited and sometimes fundamentalist sounding vocabulary toward faith practices in general. Whereas before I might have found Bloom's (1992) thesis of America as a "post-Christian nation" interesting, I now found myself a believer. The neo-Gnosticism that Bloom argues emerging in American culture generally from the influence of the Southern Baptists and Latter Day Saints was tangible for me after studying these students. At best, students of faith in general negotiated a religious identity that was lacking much content, knew little of each other, but *knew* that they believed in God, and/or knew they *were not* an atheist.

8.5 Evolution and Educational Phronesis

Adept educators who teach about evolution are quick to point out that the concept is not controversial everywhere. There are many places in the country and around the world for that matter where there appears to be no appreciable conflict. A "best practices" consensus of how to approach topics like evolution surely involves "nature of science" pedagogical techniques. But as Lederman's (2007) exhaustive review of this literature reveals, there are glaring holes in this work. Working forward may require epistemological heavy lifting of an order which few science teachers, and perhaps few science educationalists, are adequately prepared. Lederman asks "what is the influence of one's worldview on conceptions of nature of science?" If we have learned anything from religious exclusivists students, it is that the rhetoric of the science education field is attempting to move a person's ontological position.

In a recent move within the social sciences, Flyvbjerg (2001) has called to move social inquiry from an *episteme*-based model similar to that of the natural sciences, toward a *phronesis*-based model, based on the Aristotelian concept of *phronemos*. In doing so, the social sciences (of which I am including science education pedagogical research), would be strong where the natural sciences would be weak. Rather than attempt to represent the symbolic and context-dependent relations of sociocultural action in naturalistic terms, a *phronesis* or practical wisdom-based view of human action based in value rationality would reorient the social sciences. Leaving a focus on finding testable and predictable theories of human action behind, phronetic social science should ask of human social action: where are we going? Is this development desirable? Who gains and who loses, and by which mechanisms of power? What, if anything, should we do about it?

Whereas *episteme* is knowledge, *phronesis* is the action of practical wisdom. One who does the right thing in novel circumstances toward any cultural action is the *phronemos*. Bourdieu's (1977) term was the "cultural virtuoso." In his example:

> Only a virtuoso with a perfect command of his 'art of living' can play on all the resources inherent in the ambiguities and uncertainties of behavior and situation in order to produce the actions appropriate to each case, to do that of which people will say 'there was nothing else to be done', and do it the right way (p. 8).

More generally, Dreyfus (2000) has written about the *phronemos* as one who has acquired master skill at some cultural act:

> With enough experience with a variety of situations, all seen from the same perspective but requiring different tactical decisions, the competent performer seems gradually to decompose [a] class of situations into subclasses, each of which shares the same decisions, single action, or tactic. This allows an immediate response to each situation (p. 160).

Bourdieu's case from his ethnographic work involved the subtleties of gift exchange, when one knows that what had been given, without having been verbally taught, is the appropriate gift, following all of the unwritten rules of such cultural action. Likewise, the American basketball player, with just seconds left, is inbounded a pass by which the *phronemos* takes the right shot to win the game, for which everyone will have said that they had done the exactly right action, even though an array of possible right shots were possible.

In this way, what then is the right thing to do regarding evolution? This has much to do with what ends for-the-sake-of-which one serves. Is it right for the Creationist to follow a state curriculum that one *knows* is an epistemological product "of a fallen world?" Is it right for Dr. Fleischman to claim that he is "not trying to change your religion" when the epistemological foundation from which he works undermines the ontological surety of the Creationist? Is it the right thing to have belief brought into science classes in any circumstances? Likewise, one could be the *phronemos* who masterfully knows how to artfully avoid teaching evolution against all other costs, lest conflict or religious criticism manifest itself in the classroom. One might, as a teacher, take a completely secular stand and alienate religious exclusivist students who flee back from any shifting of their ontological foundation. Of the kinds of resistance I encountered and the value rationalities in the vignettes I shared, policy documents such as *Science for All Americans* (Rutherford and Ahlgren 1990) have little programmatically to say. Legal precedents such as *Kitzmiller* do little to immediately sway the social structures which produce new religious exclusivists. Teacher education training programs either completely ignore, or only superficially acknowledge, philosophical training. The sciences are often worse. Both ignore the epistemological complexity seemingly needed to get evolution education done.

8.6 Education and the Birth of Possible Worlds

A respectful social tolerance of plural religious commitments is required for sanity and peaceful coexistence in a multicultural society. But holding this line and getting evolution education done may require moving the status quo of religious discourse to a more complex and nuanced position. Evolution education requires teachers and students to deal with epistemological complexity and the role of affect in our reception of science, rather than a naïve view of one-dimensional Platonic *Truth*. Although Alsop (2005) discussed the critical need for such work in science education research, my sense is that some of the issues I have discussed as limiting more complex epistemological discussion are actually indicative of many professionals within the field.

In this work, the university discussion of evolution would likely be the first discussion of the topic for some, and in many cases likely their last. My student interviewees had previously encountered evolution in mostly limited ways, at least against the measure of the Project 2061 (AAAS) policy ideals (1993, 2001). In a step for which this project can certainly build upon, Hokayem and BouJaoude (2008) find that worldview is of utmost importance toward agreeing with evolutionary theory:

> Assuming that the various constructs of worldview…influence how individuals perceive scientific theories, identifying the relative importance of one or more of these constructs for accepting scientific theories other than evolution could provide a more profound understanding of the intricacies and problems specifically associated with the theory of evolution (p. 412).

But as Hokayem and BouJaoude's stance illustrates, about the debacle of Michael Reiss's removal from his position with the Royal Society, many of the loudest and most powerful within the natural sciences refuse to make such considerations. To do so calls into question the absolutism of their scientistic view of the world, but more pointedly requires a political deference to the interpretive tools of philosophy and the social sciences.

Some science educationalists insist on working toward a science pedagogy where beliefs are not brought into the classroom (Smith and Siegel 2004). As I have begun to articulate, this view is both epistemologically shaky and naïve toward our embeddedness in culture. Given the prime importance of how ontological positions project systems of belief and charge the discussion for all involved, ontology cannot be ignored if we are to hold better dialogues regarding evolution education.

8.6.1 Down Off the Precipice

From practical experience of talking with students and having them build up their worldview regarding evolution, I was struck. Many conceptual frameworks with which they viewed the world, I sensed, could easily be broadened if the right dialectical scene opened before them. Walking students right up to the precipice, where their religious exclusivist ontology ran out of ground, and the abyss shown below, rightly prompted anxiety. But why see an abyss? (Fig. 8.1)

Fig. 8.1 A man stands above a 1,500′ sheer drop at the edge of the Grand Canyon. While hiking nearby, simply the sight of such action elicited a wrenching of my gut

Fig. 8.2 Looking down 2,500′ into the Grand Canyon, there are ways to negotiate the abyss

Extending Swidler's (2001) hiking metaphor, even the most abyss-like situations have ways down. The experience of standing at the edge of the Grand Canyon can elicit a somewhat tangible analogue. Even though many cannot conceive of ways down (or up)—they exist (Fig. 8.2).

In our discussions of evolution, for those who object, religious ontology appears to be the sole basis from which any complaint stems. Given that the basis of science is in question when Creationist critiques of evolution emerge, a more sound pedagogy for evolution must certainly include a discussion of the historical and contingent nature of science, and the historical and contingent nature of religious faith. Advocating for a more robust historicizing of religious faith, as it stands now in US educational discourse, should surely meet with resistance from religious exclusivists. Negating such discourses partly explains the ease by which so many adhere to such absolute and seemingly naïvely arrogant views.

8.6.2 Tearing Down and Rebuilding the Scaffolding

Bruner's (1991) contribution of "scaffolding" to educational theory can articulate a way down off the precipice, if only by changing the metaphor a bit. Paralleling and in some ways explicitly bleeding into the domain of anthropology, Bruner's theorizing bears structural similarities to Geertz's "webs of significance" (Mattingly et al. 2008). Tweaking the metaphor a bit, evolution education could be moved along if we appeal to pedagogies that scaffold between each other—a means by which the significance of our worlds becomes intelligible interpersonally, and the means by which we can open new worlds of possibilities. Gopnik (2009) sets the tone for this move, in recognizing that Darwin's reorientation of the prior hierarchical view of nature, as linked to a God, had been supplanted by the now possible horizontal, historical, and evolutionary view of nature. Any move in such a direction, given the nature of ontology and commitment which we have examined, becomes a problem of "difficult dialectics" as I have discussed them (Long 2010b). There are ways of being in the world for which epistemological reflexivity is not undertaken without ontological consequence. A spirit of curiosity, questioning, and the inquiry most valued in the sciences is muted by exclusivist ontology.

Seeing science as an epistemic culture, policy ideals such as SFAA would stand to learn by employing the metaphors of anthropology. The instillation of "habits of mind" begins to sound as though one is arguing for students to "go native" to science. But this already *is* essentially the process of graduate "bench science" socialization. Naturally, the ontological commitments that I have examined clearly provide confounding examples of programmatic failure when policy ideals simply do not take culture and ideology seriously. Given the role that religion plays in structuring ontology, how can science education policy not explicitly address this? Moreover, how can this be done while remaining respective of religious pluralism?

To not account for others fully, their ontology and their *being-in-the-world*, is to make an old mistake which Nietzsche viewed of science at its worst. To make a way of life solely of the naturalistic attitude, reducing a person to a pile of elements, is a sick view of the world. Turned toward a healthier view, scientists as others are creatures of culture. Apart from powerful work of inquiry that makes up day-to-day scientific practice (and rationale that directs my interest in articulating the importance of worldview), the proposition that one continuously and fully encounters all facets of the world as "scientist" is preposterous. If we wish to welcome more into the practices of science, as I do, we might make the introduction a bit more inviting. Science at its best is an aesthetic. A view by which, with the joy of apprehending complexity, one is projected out into imagining the possibility of what might be, and what curious phenomena await beyond. Although science cannot offer transcendental *Truth*, it certainly can make our brief existence more interesting.

Chapter 9
Epilogue: How Science's Ideologues Fail Evolution or: *Richard Dawkins and the Madman*

I leave my description of evolution and culture by considering the limits of science education policy, and then consider some of the shrill discourses coloring evolution/ Creationism disagreement. Both are important to this work, but one step outside the immediate purview of the main project. As we step out of the local and contextual frame of Mason-Dixon, I will now situate these students, teachers, faculty, and community members' relationships toward evolution within more general national and international discussions.

As Toumey (2004) saw it, there are two genres when one writes about Creationism—the polemic mission and the interpretive. I now turn toward interpreting the polemicists. In a vignette that I discuss at the Stanford Aurora Forum, Richard Dawkins and astrophysicist Lawrence Krauss share an exchange that serves as an exemplar of the limits one encounters when calling for broad science literacy. As I demonstrate, when you attempt such overarching intellectual engineering, you run directly into issues of civil liberty and the oldest of questions—how should one live a life. After this analysis, I reconsider the ends that Creationists continue to work toward even in the face of Darwin's Apocalypse. Both of these views, the scientistic and the Creationist, our two rails of *Truth*, are narratives committed to a view of the world as a unified whole which each hope to describe as one unified *Truth*. Each warrant a summary comment in light of the territory I have covered.

9.1 The Ends of a Scientific World

This project was in part guided by knowledge that it might elucidate examples of cultural practice toward scientific concepts, working against science literacy, that science education policy does not appear to clearly account for. It is also likely the case that the types of distinctions I make are unstated for good reason. In these regards, I draw on AAAS *Science for All Americans* (Rutherford and Ahlgren 1990)

D.E. Long, *Evolution and Religion in American Education: An Ethnography*,
Cultural Studies of Science Education 4, DOI 10.1007/978-94-007-1808-1_9,
© Springer Science+Business Media B.V. 2011

policy as a means of illustration. On first glance, AAAS policy seems to nicely accommodate issues regarding science and culture:

> Every culture includes a somewhat different web of patterns and meanings: ways of earning a living, systems of trade and government, social roles, religions, traditions in clothing and foods and arts, expectations for behavior, attitudes toward other cultures, and beliefs and values about all of these activities (p. 151).

This view accords fairly neatly with a robust phenomenal view of culture. In fact, it is likely someone quoted Geertz as they wrote. Surprisingly, SFAA makes some relativistic statements that certainly might rankle some readers. It "treat[s] standards for defining crime and assigning punishment as cultural variables. This approach is controversial, for many people believe that there are absolute standards for acceptable and unacceptable human behavior" (p. 153). In our case, that would be anyone with a religious exclusivist ontology, and likely many of the Both/and category. This relativistic view is surprising to eyes like mine, given the following overarching foundational beliefs for SFAA. Speaking of the scientific "worldview":

> By working over time, people can in fact figure out how the world works...the universe is a unified system and knowledge gained from studying one part can often be applied to other parts....knowledge is both stable and subject to change (p. 5).

So regarding our first question, why is the treatment of "crime and punishment" a matter of "cultural variable" when the foundational claim of the nature of science is that the "universe is a unified system," for which "we can find out how the world works?" It is precisely this distinction between world and universe, so commonly slipped between even in the rigor of science policy documents, but which obfuscates distinctions of tremendous significance. Whereas the Democretian view of science as describing the interrelatedness of atoms in the void does mostly work to describe the universe (ignoring examples such as the interpretive lacunae between the Newtonian and quantum interpretations of mechanics), the significance of "world," the culturally imbued substances within them, and the self-reflective symbol making beings within it, is something that science does not quite know how to appropriately discuss. This becomes painfully clear when SFAA makes normative claims about epistemology. "New ideas are essential for the growth of science—and for human activities in general. People with closed minds miss the joy of discovery and the satisfaction of intellectual growth throughout life." What constitutes an open mind? Are Creationists proscribed from joy and satisfaction? Must one experience intellectual discovery to have a good life?

So what, given the treatment of science as a unified system where one may apprehend a totality of the world, is the basis by which the unified system rests? Profound disagreements between forms of scientism and constructivism delineate the kinds of discussions which this project will not be able to adjudicate. At the beginning of this book, I outlined the lacking ultimate foundation that Nietzsche and Kaufmann (1974) pointed toward, regarding both Western religion and science in his philosophy. The Existential/Phenomenological and American Pragmatic traditions both emphasize this. But what cannot be avoided, as this project has shown, is that absolutist epistemologies, and the religious exclusivist ontologies they are embodied in, are not amenable toward quite such an open view of the nature of science.

Socializing students toward "scientific habits of mind" is in fact a social agenda and a statement about engineering worldviews. Moving students away from an ontology which must do a fact check of scientific data with a literalistic reading of the Bible is by what it implies, an effort to change one's religion. Dr. Fleischman's comment that "we're not trying to change your religion" only superficially acknowledges what it means for a religious exclusivist to "have" a religion, ontologically. Lest one really believes that one can be an honest broker in the social practice of science education and balance old-time religion with an orthodox philosophy of scientific practice (methodological naturalism), I find this a Janus-faced misrepresentation. This novel (schizophrenic?) creature may exist, but the cases I found had a "flair" for another agenda. As a nonreligious person, and one interested in advancing the scientific literacy of our nation, as a matter of increasing the civic capacity of the republic to engage in the highest common denominators dialectic about commonly shared problems, I have problems with fundamentalist religion mediating science education. Secondly as a nonreligious person invested in evolution education, I must at least side with the Both/and position. The tension these types of discussion prompt, those of teacher "disposition," is currently hotly contested in the teacher education literature. Views such as Damon's (2005) delineate the political stakes of this debate, respective of the data I have just presented:

> It is not acceptable to assess particular attitudes and beliefs related to social / political ideologies. For example, a candidate's belief systems regarding economic redistribution, the politics of multi-culturalism, the implications of religious faith and its expression, whom we should vote for in the next election, or even whether all our national wildernesses should be turned into golf courses, are none of an assessor's business (p. 5).

Villegas (2007) represents another side of such discussion, concerned with recruiting teachers in the profession committed to social justice, and hopefully discouraging bigots from the field.

SFAA policy ideals do not directly address and really offer no clear rationale for how students are to manage the competing discourses from other social fields vying for their time. For any one student, the influences of other circles of social practice compete for the grand narrative of meaning in a student's life. In the case of Creationists, the competition is easy to spot. For the vast number of students ambivalent to science, variable interests orient the ends toward which they might apply competing "habits of mind." Family, sports, music, work, the arts, social networking in general, political organizations, and the various overarching ways sexuality interlaces into one's *being-in-the-world* all pull students in varying directions. A more difficult way to frame this is to pose the devil's advocate question— why should science win out over other "ways of knowing." Does the *Truth* of science trump other areas of significance?

One "way of knowing," that certainly could use a bit of cultural pushback is that of the more damaging rhetoric of the "new Atheists" movement. Although Richard

Dawkins has been one of the loudest in these regards lately, preeminent philosopher Bertrand Russell (1959) popularized similar views a half decade ago:

> I see no evidence whatsoever for any of the Christian dogmas. I've examined all of the stock arguments in favor of the existence of God, and none of them seem to me to be logically valid...there can't be a practical reason for not believing what isn't true...I rule it out as impossible. Either the thing is true or it isn't. If it is true you should believe it, if it isn't true you shouldn't. If you can't find out whether it is true or it isn't you should suspend judgment.

It strikes me that when I have heard these types of arguments, they have most often been stated from a position of social power, and usually out of the mouths of elder white men. In a fairly well-known similar exchange on the PBS *Charlie Rose* show (Watson et al. 2005), biologist E. O. Wilson and James Watson both explained that in their daily lives, among their closest scientific friends and colleagues, they both cannot think of anyone they know "who believes in a personal God." From this type of position, the influence of culture is seen to amount to simple preferences or tastes. It is easy to see how such privilege can tend to engender a myopic illusion of objectivity. In fact, I am almost as likely to have made such statements at points in my life. But what these types of statements, particularly the concision of Russell's, lack is a content by which most people have great interest. That which makes up and orients the narrative of one's life. Like it or not, meaning making (whether transcendental or ephemeral) is a hallmark of our current Western cultural epoch. In the extreme, as (Nietzsche and Kaufmann 1974) complained, we of the West are "meaning addicts."

Most people do not usually think about or care for that matter about issues such as biological evolution. Why in their day-to-day lives would they? This is not an antiscience statement; it is simply a description of the normative case of Americans and their relationship to science. Seeing science as needing purified of any influence from individuals who possess religious convictions, Richard Dawkins published his popularly selling *God Delusion* (2006) with the expressed intent to "kill religion." On its own, this would not have to be related to evolution education, except for the fact that Dawkins himself attributes his atheism as inspired by coming to understand natural selection as a teen. Secondly, Dawkins is a popular boogeyman for Creationists when referring to the "evils of evolution." With such narratives dominating public dialogue regarding science and public understanding, any "debate" regarding evolution tends to be essentialized as science versus religion.

As part of a larger scientistic movement, evolution tends to get caught up as an ideological playing piece. In this movement, there has been a minor explosion of in-your-face atheism books published in the last few years. Richard Dawkins, along with Christopher Hitchens, Sam Harris, and Daniel Dennett have each released a volume which fires one type of shot or another across the steadfast bow of Christian America. They do not leave out other faiths, but as Britain and America both share this dominant religious legacy, toward it goes the majority of their critique. I am pretty sure that for the most part, these shots missed their targets. I bring this up partly because the reception to these works is indicative of a more general characteristic of our civic attitude toward religion and public life. To be an atheist in most quarters of the USA is, as Edgell et al. (2006) discusses, to be the least trusted minority in American life.

Just coming out with it, to take this stand, is to hold to a position in which among mixed company puts people off. Where this position may be easy to espouse in a natural sciences department of an elite university, it is a bit trickier politically in most public school environments in the USA. Whether right or wrong, the political act of *being* an atheist, explicitly, in the current US context in most public discourse, is received as passively indicting others' belief systems in a way that equates them to a myopic caricature who insists on a "magical man in the sky." It also places the atheist, depending on the severity of who is looking, in a pitiful, vagabond, or even demonic light. As Dawkins would see it for evolution education, like Bertrand Russell, there is no need to appeal to anything other than what is clearly "reality." But, as I have made painfully clear, the phenomenon of resistance to evolution is not as simple as focusing a picture of reality for people. Whether right or wrong, wondrous or foolish, for most people, "reality" includes knowledge of a supernatural force, most often a God. Whether it should be set aside, religious belief is a normative value for a majority of Americans.

I do not disagree with most of what Dawkins, Dennett, Harris, and Hitchens have to say. What I do take issue with are parts of the approach that Dawkins takes, and the effect it has on the mission of evolution education. For a just relinquished Oxford Chair of the Public Understanding of Science, whose parallel interest is to "kill religion," there are some issues for which he has failed his mission. Consider his lacking understanding of the context of religious practice and how it bears on science education in the USA. Speaking to Dawkins at the Stanford Aurora Forum, astrophysicist Lawrence Kruass details where he thinks he has had the most success in reaching out to Creationists:

> Krauss: I think that the largest impact I've ever had is going to fundamentalist colleges and saying just the simple statement that you don't have to be an atheist to believe in evolution. I've had kids come up to me and say, "I've never heard anyone say that."
>
> Dawkins: How could they possibly not have heard anyone say that? Haven't they ever talked to a clergyman, a bishop, an archbishop?
>
> Krauss: No, you see this isn't the Church of England…These kids, every Sunday, from the time they're too young to think—and that's why both you and I absolutely agree that it's sort of child abuse to subject children to that—but they are told that. They are told that in this country, and that is why many of them just have this gut reaction (Stanford University 2008).

No Richard, it is certainly not the Church of England, and most religious people likely have very few conversations regarding science as part of their liturgy. But countering this, as I do know, there are many congregations, in fact entire denominations, that take an active stand against evolution as part of their services. It is also likely that, although smaller in number, this same message is present in some British congregations. As Numbers (2006) sees it, this is also set to grow throughout the world.

This focus on religion is unavoidable when discussing evolution. If Both/and and agnostic/atheist teachers do not, others with a "flair" for teaching will. This is not because thinking about evolution purely unto itself requires one to consider teleological thoughts; but rather, most students and likely many teachers come to consider science generally and evolution in particular from an unexamined teleological position. In drastically different circumstances than Oxford or Stanford, everyday evolution education takes place in the whole of one's being—in schools, universities, homes, among friends, and in places of worship. This is where the hard work of evolution education takes place, and this is where science and science educators have largely failed against the call of competing narratives.

Part of what is muddying the issue for scientists in the public eye is the lack of a clear picture of the sociology of antievolutionary attitudes, and how this plays out in educational practice. Returning to Krauss's discussion of his success in interpreting evolution at fundamentalist colleges, this is the type of public outreach and educational stance that we can use and be proud of in the USA. But quickly failing this, Krauss shows his own disconnect between his experience and his perception of educational norms:

> My daughter went to a rather liberal private school, but her teacher, and this is where [Creationists] have been really successful: even if they haven't distorted the curriculum, they've instilled enough fear in teachers that there are many teachers in this country, biology teachers, who won't mention the word evolution, not because it's not there, but they are just afraid. Even in my daughter's school, I remember her biology teacher basically apologized each time he mentioned the word evolution. I was just shocked, but it happens.

And from his personal teaching experience among the "norm":

> I taught at Yale University for what seemed like an eternity, and I learned that you can be very successful in this world not only just being scientifically illiterate but proudly proclaiming it, and it's unfortunate. It really is.

Now if we cannot get this done at Yale, then some might say we might as well give up. In fact, to turn to *Yale* as some sort of norm is just silly, but it also shows the ways in which the most prominent of our scientific voices may be just a bit out of touch with the norms of the educational system. Add to this Creationist Michael Behe's telling statements about his commitment to public education, and it becomes clear that there is clearly a lacking understanding of the social practices which normally typify evolution education. In an "extras" scene on the DVD presentation of Mark Olsen's (Calrlisle and Olsen 2007) film *Flock of Dodos*, Behe, who testifies in *Kitzmiller*, potentially affecting public school curricular policy, dismisses any tangible commitment to public education, by dismissively conceding that his own children do not attend public schools.

In addition to the anti-evolutionary fringe, science literacy advocates must contend with the broad disconnected and uninterested masses. But, unlike Dawkins and Krauss, are we *really still surprised* at the state of American science literacy? The famous science education video *A Private Universe* in which Harvard graduates fumble through basic facts about the solar system already made this point (Private Universe Project 1989). But as I have endeavored to show, the socialization of scientists, how some of them view the world, and the practicalities and social capital of their jobs are as much at fault in this debate as any other factor. What usually gets left out of the science literacy picture, with the exception of a few studies, is the importance of the social worlds from where students originate, and how this influences their interest, epistemological framework, and receptivity toward science when they encounter most science classrooms.

There is an important issue that gets glossed over in most discussions regarding evolution and science education. SFAA gives many examples of rational warrants for which science education should work. An improved workforce for market competitiveness, positioning one's self in this milieu, or preparing for future global conflicts, each of which are held out as social or individual goods. But as Cobern points out (2007), buried behind and underpinning this social agenda (and supported by the likes of Dawkins, Dennett, et al.) is the presumption that metaphysical naturalism is both the way to go for science, and the "only game in town," not only for looking at the natural world, but for *all human value*. The error that Dawkins makes is that in leaping from methodological naturalism to metaphysical naturalism, he has in some way shown a warrant that this is *the* pathway toward some better form of life. In his book *A Devil's Chaplain*, (Dawkins and Menon 2003) is especially dismissive about attempts by social scientists to contextualize and illustrate the lack of supposed scientific objectivity in normal scientific practice. As a general comment on his gripe with social constructivists, he claims that constructivists (while riding in airplanes) are not constructivists at 30,000 ft. Harry Collins (1995), pointing the same lack of understanding that he sees in Dawkins, asserts: "he has money in his pocket in the friendly skies…and money is socially constructed." Dawkins can show as many examples of lay trust in the operations of airplanes or in contemporary medicine as he likes, but most people stand back puzzled at this distinction. For to make this move, to insist that metaphysical naturalism is somehow clearly *the* move to make, requires that people value science in the ways that he does. For most students and many scientists alike, they clearly do not. Why would we ever think that they would?

People want meaning in life. People make meaning in life. Some meaning is profound, some is silly. Most people believe that meaning is sent down by a God or gods. Science does not, at least in the conventional sense, give us meaning. Science, à la Hume, gives us a description of the natural world. As much as some scientists continue to mess this point up, science does not "progress" as a universal metric unto itself (Nitecki 1988). Is a more and more detailed description progress? It depends on what ends you serve. That the *limulidae* (horseshoe crabs), having succeeded in "progressing" through vast stretches of geologic time unchanged, measures against human "progress" in what way? By itself, data produced by measuring and watching behaviors of these crabs have no way to bear on any conception of progress. Now there are problems for which we can come to agree to apply appropriate means, but any subsumission of other human meaning by science is both on very philosophically questionable ground, and more importantly, to what ends? In more eloquent terms, this hybridization of various forms of *episteme* mediated by social arrangements is, in quite an (ironically for Dawkins) evolutionary way, the thesis that Latour (1993) works in *We have never been modern*. Variously equipped and committed to ways of being informed by both naturalistic and cultural styles, we do, as a matter of social epistemology, evolve in our epistemological style as culture tacks back and forth reflexively with narratives about nature.

Exemplifying the specific form of naiveté exhibited by those who refuse to see social construction of knowledge, Nietzsche's (Nietzsche and Kaufmann 1974) aphorism of the Madman still speaks to those, such as Dawkins, directly. For those like Dawkins who would disembowel our culture's dearest belief structures, without offering salient alternatives:

> God is dead. God remains dead. And we have killed him. How shall we comfort ourselves, the murderers of all murderers? What was holiest and mightiest of all that the world has yet owned has bled to death under our knives: who will wipe this blood off us? What water is there for us to clean ourselves? What festivals of atonement, what sacred games shall we have to invent? Is not the greatness of this deed too great for us? Must we ourselves not become gods simply to appear worthy of it? (p. 181).

So, with blood on his hands, Dawkins stands with the blade of his *God Delusion* (2006). This same educator of the "public understanding of science," when challenged, often resorts to *ad-homonym* assessments of people from different perspectives as being "stupid or thick" (Stanford University 2008, p. 27). Now do not get me wrong, I also have an aversion to scientifically illiterate public officials making decisions which affect social issues (as Dawkins was referencing). But quite unlike the *de fatco* stance Dawkins takes, I am actually interested in the ways that people would come to think differently than me, and how that knowledge can be used to shape better educational strategies.

Before we let Dawkins off the hook, let us consider the issue of "world" one last time. There are many, many "worlds" for which Dawkins would appear "stupid or thick." Perhaps if we quizzed him on the genres and issues within recent popular music or on the issues at stake in American college basketball, both socially "real" phenomena" of which many have great interest, he might come up lacking. It might work better to put him in a scenario in which he was publicly embarrassed about such matters. But being "stupid or thick" about these worlds of competing human interest does not negate the meaning that they have or take away popular commitments to them. As I would expect, he might harrumph at the prospect of being pinned down to such tasks. But this is precisely the issue at work in our failure to improve science literacy and dissolve aversion to evolution. Why, given the content of people's everyday lives and understanding of the world, should they care?

The conundrum this raises is well exemplified in the same Stanford public forum visited above. After concluding their conversation, the very first question from the audience demonstrates the nature of what I am getting at:

> I'm very thankful for you gentlemen—for what you do. I wasted a great portion of my life trying to find purpose in the universe. But I do have a quibble with you, and that is with respect to awe and wonder. Awe and wonder, I believe were the greatest tools in the origin, propagation, and promotion of religion and I don't think that...I know in religion they want to emphasize awe—that's the reason they built cathedrals—that's the reason they monopolized all forms of entertainment during the middle ages. Today we have the most boring entertainment you can imagine on television on many channels every day, and the public is going to eventually get sick of this I think. When we teach science I think it should be factual, because if there is a purpose, I think it is to survive...and to survive, we need to know the facts. And I'm very tired of constantly trying to whoop up entertainment and emotions. Emotions were, in the origins of the universe a survival mechanism for animals...for us, emotions are a destructive mechanism. So I hope that you can tell people...defend science without having to tell them that it's awesome—*IT'S REAL!*" (Stanford University 2008).

Krauss, with a nod from Dawkins, vehemently disagreed with the proposition, and took a noble stand for the value of scientific curiosity and its products as ends unto themselves. Preceding this question, Dawkins makes the distinction between the "awe and wonder" value of science versus the "nonstick frying pan" value— goodies collected as instrumental side effects of the space exploration industry. This instrumentalism is something that both Krauss and Dawkins see as a lacking of significant enough warrant for broad science literacy. But this example does raise the *sine qua non* for scientists in such positions. What purely rational warrant do they produce to charm the public toward science? Awe, beauty, grandeur, mystery— are each of themselves moods and/or aesthetics of human existence. There is no absolute scientific measure of grandeur, awe, or beauty, because by scientific terms, *they are not real!* As you saw during this project, anxiety and revulsion can be added to the types of existential phenomena people encounter when contemplating complex scientific concepts.

Pointing out this lacking discourse from scientists is easy. It is not their fault that some of them sound this way. They are not trained or charged with providing human society with social hope or the direction of purpose. At the same time, some still scratch their heads when students offer up teleologically framed conceptions in their answers. How, for the Creationist and the vast amount of half-interested religiously grounded students, are they to be expected to set aside their ontology while doing science? This is not lost on more astute theological scholars critical of what they see as a curious silence from the *Oracles of Science* (Giberson and Artigas 2007). As Giberson and Artigas see it, preeminent champions of the grand evolutionary narrative like Stephen Jay Gould, Richard Dawkins, Steven Weinberg, E.O. Wilson, and Carl Sagan leave an entirely unsatisfactory taste in the broad mouth of culture. Critiquing each of their answers to how evolution as grand narrative provides meaning to human experience:

> Gould simply denies it, arguing that the process that created us is just a series of accidents, no more a source of meaning than a gambling casino, where some people strike it rich. Dawkins, the Devil's chaplain, thinks the evolutionary epic is a grand and meaningful tale compared to the pitiful analogue in religion, but he certainly offers no suggestions for how ordinary people, uninterested in the creation story of science, should make their way in the world. Sagan's view is more positive but sometimes seems to rest on the speculation that we will one day establish a connection with extraterrestrial intelligent beings....Weinberg is pessimistic about the quest for meaning, seeing human experience as farce. Wilson is the more optimistic of the Oracles, but his optimism relies on developing an understanding of the biological basis of human behavior that will prove adequate to provide meaning....The Oracles have all challenged traditional religion in various ways, despite their failure to produce viable replacements (p. 231).

In the end, I am on Krauss and Dawkin's side, but coming from another direction respective of Giberson and Artigas. The complexity and nuance demonstrable and illustrated by science is what might sustain our commitment to it as an epistemology in practice over the long term. But this holistic appreciation of science is not part of the usual official discourse of why one does science. As we currently practice it both in the K-12 and university settings, if a student comes to see science this way, it is of their own volition, interest, or philosophic synthesis. It is not due to explicit discussion. Perhaps we would be better off considering such discussion.

9.2   The End(s) of a Creationist World

As Creationist students expressed, the ends they saw for their lives were set forth by Biblical edict. Tyson, speaking about discussing his growing skepticism at what he does and does not believe, talked about the anxiety of "not knowing why he should be good in the first place" at the prospect of his belief system being displaced. For many students, the singularly loud call of their exclusivist faith made the prospect of questioning it unlikely. As Wuthnow (2005) best described it, "Jesus becomes so attractive to the exclusive Christian that he or she finds it difficult to understand why anyone who has heard about Jesus—and most people have, they presume—would not immediately become a Christian" (p. 176). As we saw in the cases of Cindy, Renee, and Nolan, it was only through major life upheaval that a window opened up within which ontological positionality toward evolution moved.

With this ideology of Biblical *Truth*, the increasingly diverse social milieu of American cultural life actually exacerbates the extremes of exclusivist Christians. Immediately following President Barack Obama's first State of the Union Address, Ken Ham and Answers in Genesis held their own "State of the Nation II" speech in response to what they see as a further erosion of American society by turning away from Biblical authority. Focusing on President Barack Obama's mention that the USA was no longer just a Christian nation, Ham points his analysis to the influence of an evolutionary worldview as we have seen. As Wuthnow (2005) describes, and is worth quoting again:

> It is not uncommon for exclusive Christians to associate the growth in diversity with some frightening, apocalyptic vision of the end times. The threat is more likely to be cast in terms of homosexuality, promiscuous lifestyles, or relativistic values being taught in public schools, any of which may be loosely associated in people's minds with diversity (p. 184)

Uncannily, this message is the corpus of Answer in Genesis's ministry. Its Creation Museum emphasizes this narrative. This same foundational message underwrote the homeschooling literature and organizational principles I examined during this project.

As Intelligent Design advocate Philip Johnson famously quipped in the infamous Wedge Document proffered by the Discovery Institute, uncovered during the *Kitzmiller v. Dover Trial* of 2005, the message is not really about science. It is squarely about cultural takeover:

> Our strategy has been to change the subject a bit so that we can get the issue of intelligent design, which really means the reality of God, before the academic world and into the schools. This isn't really, and never has been a debate about science. It's about religion and philosophy (Johnson quoted in Forrest 2007).

What evolution portends for Creationism is the cultural death of a worldview, and a way of *being-in-the-world*. This is not lost on them, and thus the likely focus on ultimate foundations and the importance of a "plain reading" of Biblical *Truth*, as Ken Ham often describes it.

As paleontologists fill in a picture of the fossil record, the "missing link" metaphor will always follow these discussions. Even referring to intermediary species is part of a conceptual trap which emphasizes any one specie's existence at any one time as an essentialized kind. Similar to the conceptual infinite in Hilbert's Grand Hotel paradox of set theory, Creationists will be infinitely able to reject evolution as each presented fossil will be seen as an essentialized iteration of some essential kind, having missed the point of the mutability of species entirely. In these regards, debating the inaccuracies or fanciful extremes of their Creation Museum's interpretive structure for its details is a waste of time. But closely watching them for their social impact is not.

Older Creationist arguments from the movement's foundational days of Henry Morris attempted to point out that evolution as a general process violated the second law of thermodynamics—that natural systems tend to break down into more disorder. This ignores the input of energy within biological systems (for us by eating). But the argument itself is metaphorically illustrative of the ends of a Creationist worldview. Seeing themselves as literally ascending Jacob's ladder, their ontology and epistemological surety cannot be derailed. Evolution attempts then to switch out the gilded rungs of their ascent with a grindingly incessant reality of biological life. Like walking up a down escalator, we do successfully fight off entropy, but only until death. The implication of evolution, that we will never get to the top of the escalator, is both what was missing in Herbert Spencer's socially implicated read into evolution and is that which the commonplace understanding of it is still grounded in. Teaching about philosophical implications of evolution in a nation that has divine providence interlaced in its mythos will always run into problems. Student conceptions of evolution almost always involve language echoing Spenserian hierarchy: better, improved, and more complex. Although complexity certainly exists, one tends to confuse complexity as a product of process with teleology, as Gould discusses (2002).

As a matter of purpose, or along one final designated path, evolution is not going anywhere. Our existence, although interesting and perhaps extremely novel, does not negate or diminish the curious existence of other species around us. Those that have come before us, and those that have yet to come are equally compelling. Because no species *arrives* at any place, this is perhaps the most deflating realization that many make when comprehending evolution. This is also the root of our inability to remove teleological language from our discussions of science. *Why* one sees design, purpose, and order in the universe, or one lacking transcendence could be the more interesting conversation we might have. Doing this broadly in civic discourse with ecumenical intent, given the theological and scientific literacy I encountered is likely a long way off.

9.3   Evolution Education in Possible Worlds

How does the preceding story relate to the educational road ahead? Critiquing a popular rhetorical and epistemological strategy for its limitations, we broaden our purview to more seriously account for education as a process that occurs embedded in culture. With the publication of the non-overlapping magisterium argument, Gould (1999) hoped to describe a comfortable epistemological position by which science and religion might be seen to be asking and providing answers to and for completely different lines of thinking. In his preamble:

> I speak of the supposed conflict between science and religion, a debate that exists only in people's minds and social practices, not in the logic and proper utility of these entirely different and equally vital subjects. I present nothing original in presenting the basic thesis; for my argument follows a strong consensus accepted for decades by leading scientific and religious thinkers alike (p. 3).

The intervening years saw retorts from Dawkins (1998), which pointed out the obvious problem within the logic of Gould's claim:

> It is completely unrealistic to claim, as Gould and many others do, that religion keeps itself away from science's turf, restricting itself to morals and values. A universe with a supernatural presence would be a fundamentally and qualitatively different kind of universe from one without. The difference is, inescapably, a scientific difference. Religions make existence claims, and this means scientific claims. The same is true of many of the major doctrines of the Roman Catholic Church. The Virgin Birth, the bodily Assumption of the Blessed Virgin Mary, the Resurrection of Jesus, the survival of our own souls after death: these are all claims of a clearly scientific nature. Either Jesus had a corporeal father or he didn't. This is not a question of "values" or "morals"; it is a question of sober fact. We may not have the evidence to answer it, but it is a scientific question, nevertheless. You may be sure that, if any evidence supporting the claim were discovered, the Vatican would not be reticent in promoting it.

Perhaps one oddity of the two views just presented is that they come from two nonreligious people—one who professed a skeptical agnosticism and one mentioned earlier whose current intellectual project is an attempt to argue for "killing religion."

How then, does Reiss's (2009) claim that addressing differing worldviews, damage the project of teaching classroom science rather than assist? As he puts it and I have been demonstrating, "a student who believes in creationism can be seen as inhabiting a non-scientific worldview, that is, a very different way of seeing the world. One very rarely changes one's worldview as a result of a 50 min lesson, however well taught" (p. 100).

As I have begun to demonstrate through the cultural logics of Creationists and other student lives, the issue of a person's ontological stance is critical to working with the concept of evolution. Ontological exclusivist views of the world and their *Truth* are the sole motivating basis by which Creationism pushes forward. Although Gould claims above that "a debate...exists only in people's minds and social practices," his first mistake is to deflate the importance of this distinction. Social practices make up the intelligibility of one's world, by which epistemic commitments become possible. His second mistake, "the logic and proper utility of these entirely different and equally vital subjects," is to draw on an illusory understanding of logic, for which our day-to-day lives simply do not correspond, and to fall victim to any conception of "*proper* utility." In quite evolutionary ways, people make and remake the semiotic significance of their worlds and their employment of the entities within them.

A robust evolution education stands to gain by explicitly historicizing and philosophically orienting students. As social practices toward evolution currently stand, our broad-based lacking repertoire of theological knowledge is as damaging toward evolution education as poor science education itself. Ironically, the multicultural perspectives that have been ushered into educational norms have extended a tolerance of "all the theories," ultimately and ironically bringing Creationists to the table out of an ethos of inclusivity. When Both/and students become voting and child-rearing adults, their anxiety at alienating no one subversively gives tremendous social power to Creationists.

Explicitly teaching more students not only the nature and limits of science, but the nature and limits of religious faith, from their mutual roots in philosophy would at least enrich our ability to negotiate difference. We must start with the training of our future science teachers, but this is likely no less an issue for training in the sciences in general. Projecting such conversations cannot predict the dissolving of Creationism, but would at least make the educational discourse regarding evolution more interesting.

References

Aerts, D., Apostel, L., De Moor, B., Hellemans, S., Maex, E., Van Belle, H., & Van Der Veken, J. (1994). *Worldviews, from fragmentation towards integration*. Brussels: VUB Press.

Affanato, R. E. (1986). *A survey of biology teachers' opinions about the teaching of evolution and/ or the creation model in the United States in public and private schools*. Ph.D. thesis, University of Iowa, Iowa City.

Ahuja, A. (2009) Evolution is God's work - Michael Reiss. *The Times*. Accessed online at: http://www.timesonline.co.uk/tol/news/science/article6425138.ece

Ainsworth-Land, G. T., & Jarman, B. (2000). *Breakpoint and beyond: Mastering the future—today*. Champaign, Ill: Harper Business.

Alexakos, K. (2010). Religion, nature, science education and the epistemology of dialectics. *Cultural Studies of Science Education, 5*(1), 237–242.

Alsop, S. (2005). *Beyond Cartesian dualism: Encountering affect in the teaching and learning of science. Science & technology education library* (Vol. 29). Dordrecht: Springer.

Apple, M. W. (1993). *Official knowledge: Democratic education in a conservative age*. New York: Routledge.

Apple, M. W. (2006). *Educating the "right" way: Markets, standards, God, and inequality*. New York: Routledge.

Bergman, J. (1979). The attitude of university students toward the teaching of creation and evolution in the schools. *Origins, 6*, 60–70.

Bernstein, B. B. (1971). *Theoretical studies towards a sociology of language. Class, codes and control* (Vol. 1). London: Routledge & K. Paul.

Bieber, J. (1999). Cultural capital as an interpretive framework for faculty life. In J. C. Smart (Ed.), *Higher education: Handbook of theory and research* (Vol. XIV). New York: Agathon.

Bilica, K. L. (2001). *Factors which influence Texas biology teachers' decisions to emphasize fundamental concepts of evolution*. Ph.D. thesis, Texas Tech University, Lubbock.

Blattner, W. D. (2006). *Heidegger's being and time: A reader's guide. Continuum reader's guides*. London: Continuum.

Bloom, H. (1992). *The American religion: The emergence of the post-Christian nation*. New York: Simon & Schuster.

Bourdieu, P. (1977). *Outline of a theory of practice* (Cambridge studies in social anthropology, Vol. 16). Cambridge: Cambridge University Press.

Bourdieu, P. (1984). *Distinction: A social critique of the judgment of taste*. Cambridge: Harvard University Press.

Bourdieu, P. (1998). *Practical reason: On the theory of action*. Stanford: Stanford University Press.

Bourdieu, P., Passeron, J.-C., Nice, R., & Bottomore, T. (1977). *Reproduction in education, society and culture/Transl. from the French by Richard Nice; with a foreword by Tom. Bottomore*. London: Sage.

D.E. Long, *Evolution and Religion in American Education: An Ethnography*,   181
Cultural Studies of Science Education 4, DOI 10.1007/978-94-007-1808-1,
© Springer Science+Business Media B.V. 2011

Bruner, J. (1991). The narrative construction of reality. *Critical Inquiry, 18*(1), 1–21.

Carlisle, T., & Olson, R. (2007). *Flock of dodos. The evolution and intelligent design circus.* New York: Docurama Films.

Chapman, M. (2001). *Trials of the monkey: An accidental memoir.* New York: Picador.

Clough, E. E., & Wood-Robinson, C. (1985a). Children's understanding of inheritance. *Journal of Biological Education, 19*(4), 304–310.

Clough, E. E., & Wood-Robinson, C. (1985b). How secondary students interpret instances of biological adaptation. *Journal of Biological Education, 19*(2), 125–130.

Cobern, W. W. (1996). Worldview theory and conceptual change in science education. *Science Education, 80*(5), 579–610.

Cobern, W. W. (2000). The nature of science and the role of knowledge and belief. *Science Education, 9*, 3.

Cobern, W. W. (2007). ID hysteria says more about some people's Freudian "ID" than about science. *Canadian Journal of Science, Mathematics, and Technology Education, 7*(2/3), 257–262.

Collins, H. (1995). Being and becoming [A review of *Searle 1995*]. *Nature, 376* (13).

Collins, F. S. (2006). *The language of God: A scientist presents evidence for belief.* New York: Free Press.

Damon, W. (2005). Fwd: Personality test: The dispositional dispute in teacher preparation today, and what to do about it. *Arresting Insights in Education, 2*(3), 1–6.

Dawkins, R. (1998). When religion steps on science's turf. *Free Inquiry.* Retrieved from http://www.secularhumanism.org/library/fi/dawkins_18_2.html

Dawkins, R. (2006). *The God delusion.* Boston: Houghton Mifflin.

Dawkins, R., & Menon, L. (2003). *A devil's chaplain: Selected essays.* London: Weidenfeld & Nicolson.

Dennett, D. C. (2006). *Breaking the spell: Religion as a natural phenomenon.* New York: Viking.

Dickerson, D. (2003). *Understanding the relationship between science and faith, the nature of science, and controversial content understandings.* Ph.D. thesis, North Carolina State University, Raleigh.

Discovery Institute. (1999). *The wedge.* Discovery Institute: Center for renewal of science and culture. Retrieved from http://www.antievolution.org/features/wedge.pdf

Dreyfus, H. L. (1972). *What computers can't do; A critique of artificial reason.* New York: Harper & Row.

Dreyfus, H. L. (1991). *Being-in-the-world: A commentary on Heidegger's Being and Time, division I.* Cambridge: MIT Press.

Dreyfus, H. (2000). Reinterpreting division I of Being and Time in the light of division II. In J. E. Faulconer & M. A. Wrathall (Eds.), *Appropriating Heidegger.* Cambridge: Cambridge University Press.

Dreyfus, H. L. (2005). Foreword. In C. J. White & M. Ralkowski (Eds.), *Time and death: Heidegger's analysis of finitude.* Aldershot: Ashgate.

Ecklund, E. H. (2010). *Science vs. religion: What scientists really think.* New York: Oxford University Press.

Edgell, P., Gerteis, J., & Hartmann, D. (2006). Atheists as "other": Moral boundaries and cultural membership in American society. *American Sociological Review, 71*(2), 211–234.

Eglin, P. G. (1984). *Creationism vs. evolution: A study of the opinions of Georgia science teachers.* Ph.D. thesis, Georgia State University, Atlanta.

Ervin, J. A. (2003). *Effects of student ontological position on cognition of human origins.* Ph.D. thesis, Ohio State University, Columbus.

Fensham, P. J. (2009). The link between policy and practice in science education: The role of research. *Science Education, 93*(6), 1076–1095.

Feyerabend, P. (1987). *Farewell to reason.* London: Verso.

Fish, S. (2005). Briefings—Academic cross dressing. How intelligent design gets its arguments from the left. *Harper's.* 70.

Flyvbjerg, B. (2001). *Making social science matter: Why social inquiry fails and how it can succeed again.* Oxford: Cambridge University Press.

Flyvbjerg, B. (2006). Five misunderstandings about case-study research. *Qualitative Inquiry, 12*(2), 219–245.

Forrest, B. (2007). *Understanding the intelligent design creationist movement: Its true nature and goals.* Center for inquiry position paper. Retrieved from http://www.centerforinquiry.net/uploads/attachments/intelligent-design.pdf

Forrest, B., & Gross, P. R. (2004). *Creationism's Trojan horse: The wedge of intelligent design.* Oxford: Oxford University Press.

Fuerst, P. A. (1984). University student understanding of evolutionary biology's place in the creation/evolution controversy. *The Ohio Journal of Science, 84*(5), 218–228.

Garfinkel, H. (1960). *Conditions of successful degradation ceremonies.* Indianapolis: Bobbs-Merrill, College Division.

Gauch, H. (2009). Science, worldviews and education. In M. Matthews (Ed.), *Science, worldviews and education* (Reprinted from the Journal *Science & Education, 18*(6–7)). Dordrecht: Springer.

Geertz, C. (1973). *The interpretation of cultures; selected essays.* New York: Basic Books.

Giberson, K., & Artigas, M. (2007). *Oracles of science: Celebrity scientists versus God and religion.* Oxford: Oxford University Press.

Goffman, E. (1959). *The presentation of self in everyday life.* Garden City: Doubleday.

Gopnik, A. (2009). *Angels and ages: A short book about Darwin, Lincoln, and modern life.* New York: Alfred A. Knopf.

Gould, S. J. (1999). *Rocks of ages: Science and religion in the fullness of life. The library of contemporary thought.* New York: Ballantine Pub. Group.

Gould, S. J. (2002). *The structure of evolutionary theory.* Cambridge: Belknap Press of Harvard University Press.

Gross, P. R., Goodenough, U., & Finn, C. E. (2005). *The state of state science standards.* Washington, DC: Thomas B. Fordham Institute.

Habermas, J. (1989). *Theory of communicative action.* Boston: Beacon Press.

Hahn, D. (2005). *Social, moral, and temporal qualities: Pre-service teachers' considerations of evolution and creation.* Ph.D. thesis, Arizona State University, Tempe.

Hallden, O. (1988). The evolution of the species: Pupil perspectives and school perspectives. *International Journal of Science Education, 10*(5), 541–552.

Harris, S. (2006). *Letter to a Christian nation.* New York: Knopf.

Hassard, J. (2005). *The art of teaching science inquiry and innovation in Middle School and High School.* New York: Oxford University Press.

Heidegger, M. (1962 [1927]). *Being and time.* New York: Harper.

Hitchens, C. (2007). *God is not great: How religion poisons everything.* New York: Twelve.

Hokayem, H., & BouJaoude, S. (2008). College students perceptions of the theory of evolution. *Journal of Research in Science Teaching, 45*(4), 395–419.

Husserl, E., & Welton, D. (1999). *The essential Husserl: Basic writings in transcendental phenomenology. Studies in Continental thought.* Bloomington: Indiana University Press.

Ingersoll, R. (2006). Turnover among mathematics and science teachers in the U.S. In J. Rhoton & P. Shane (Eds.), *Teaching science in the 21st century.* Arlington: NSTA Press.

Jensen, J. M. (1999). Creating a continuum: An anthropology of postcompulsory education. *Anthropology & Education Quarterly, 30*(4), 446.

Jorstad, S. (2002). *An analysis of factors influencing the teaching of evolution and creation by Arizona high school biology teachers.* Ph.D. thesis, University of Arizona, Sun City.

Kierkegaard, S. (1957). *The concept of dread.* Princeton: Princeton University Press.

Kierkegaard, S. (1962). *The present age: And of the difference between a genius and an apostle.* New York: Harper & Row.

Kimball, M. S. (2009). *Empirics on the origins of preferences: the case of college major and religiosity* (NBER Working Paper, No. 15182). Cambridge: National Bureau of Economic Research.

Kincheloe, J. L. (2008). *Knowledge and critical pedagogy: An introduction* (Explorations of educational purpose, Vol. 1). Montreal: Springer.

Kincheloe, J. L., & Tobin, K. (2009). The much exaggerated death of positivism. *Cultural Studies of Science Education, 4*(3), 513–528.

Kliebard, H. M. (1987). *The struggle for the American curriculum, 1893–1958*. New York: Routledge & Kegan Paul.

Kuhn, T. S. (1970). *The structure of scientific revolutions*. Chicago: University of Chicago Press.

Kyzer, P. M. (2004). *Three Southern high school biology teachers' perspectives on teaching evolution: Sociocultural influences*. Ph.D. thesis, University of Alabama, Tuscaloosa.

Labaree, D. F. (1997). Public goods, private goods: The American struggle over educational goals. *American Educational Research Journal, 34*(1), 39–81.

LaHaye, T. F., & Noebel, D. A. (2000). *Mind siege: The battle for truth in the new millennium*. Nashville: Word Pub.

Lakoff, G., & Johnson, M. (1980). *Metaphors we live by*. Chicago: University of Chicago Press.

Lareau, A. (2003). *Unequal childhoods: Class, race, and family life*. Berkeley: University of California Press.

Larson, E. J. (1985). *Trial and error: The American controversy over creation and evolution*. New York: Oxford University Press.

Latour, B. (1993). *We have never been modern*. Cambridge: Harvard University Press.

Lear, J. (2006). *Radical hope: Ethics in the face of cultural devastation*. Cambridge: Harvard University Press.

Lederman, N. (2007). Nature of science: Past, present, and future. In S. K. Abell & N. G. Lederman (Eds.), *Handbook of research on science education* (pp. 831–879). Mahwah: Lawrence Erlbaum Associates.

Lerner, L. S. (2000). *Good science, bad science: Teaching evolution in the states*. Washington, DC: Thomas B. Fordham Foundation.

Levine, A. (2006). *Educating school teachers*. Washington, DC: Education Schools Project.

Long, D. E. (2010a). Scientists at play in a field of the Lord. *Cultural Studies of Science Education, 5*(1), 213–235.

Long, D. E. (2010b). Science, religion and difficult dialectics. *Cultural Studies of Science Education, 5*(1), 257–261.

Lovan, D. (2010). *Top home-school text dismisses Darwin, evolution*. Associated Press. Retrieved from http://news.yahoo.com/s/ap/us_rel_home_school_evolution/print

Lucas, A. M. (1971). The teaching of "adaptation". *Journal of Biological Education, 5*(2), 86–90.

Lynd, R. S., & Lynd, H. M. (1929). *Middletown. A study in contemporary American culture, etc.* London: Constable & Co. printed in U.S.A.

Marty, M. E. (1998). Revising the map of American religion. *The Annals of the American Academy of Political and Social Science, 558*, 13–27.

Mattingly, C., Lutkehaus, N. C., & Throop, C. J. (2008). Bruner's search for meaning: A conversation between psychology and anthropology. *Ethos, 36*(1), 1–28.

Merleau-Ponty, M. (1962). *Phenomenology of perception. International library of philosophy and scientific method*. New York: Humanities Press.

Miller, K. R. (1999). *Finding Darwin's God: A scientist's search for common ground between God and evolution*. New York: Cliff Street Books.

Miller, K. R. (2008). *Only a theory: Evolution and the battle for America's soul*. New York: Viking Penguin.

Mooney, C. (2005). *The Republican war on science*. New York: Basic Books.

Moore, R. (2007). The differing perceptions of teachers & students regarding teachers' emphasis on evaluation in high school biology classrooms. *The American Biology Teacher, 69*(5), 268–271.

Morris, H. M. (1974). *Scientific creationism*. San Diego: Creation-Life.

Naugle, D. K. (2002). *Worldview: The history of a concept*. Grand Rapids: W.B. Eerdmans Pub.

Nietzsche, F. W., & Kaufmann, W. A. (1974). *The gay science; With a prelude in rhymes and an appendix of songs*. New York: Vintage Books.

Nitecki, M. H. (1988). *Evolutionary progress*. Chicago: University of Chicago Press.

North, J. B. (1994). *Union in truth: An interpretive history of the Restoration movement*. Cincinnati: Standard Pub.

Numbers, R. L. (1998). *Darwinism comes to America*. Cambridge: Harvard University Press.

Numbers, R. L. (2006). *The creationists: From scientific creationism to intelligent design*. Cambridge: Harvard University Press.

Overton, W. (1982). Creationism in schools: The decision in McLean versus the Arkansas Board of Education. *Science, 215*(4535), 934–943.

Patterson, R. T. (2006). *Evolution exposed: Your evolution answer book for the classroom*. Hebron: Answers in Genesis.

Pew Forum on Religion & Public Life. (2010). *US religious knowledge survey*. Washington, DC: Pew Forum on Religion & Public Life.

Pew Foundation (2008). *U.S. religious landscape survey*. Retrieved from http://religions.pewforum. org/pdf/report-religious-landscape-study-full.pdf

Pew Research Center (2005). *Reading the polls on evolution*. Retrieved from http://people-press. org/commentary/display.php3

Plenty-coups, & Linderman, F. B. (2003). *Plenty-coups, chief of the Crows: With a new, previously unpublished essay by the author*. Lincoln: University of Nebraska Press.

Private Universe Project. (1989). *A private universe [Videotape]*. Cambridge: Harvard-Smithsonian Center for Astrophysics.

Project 2061 (American Association for the Advancement of Science). (2001). *Atlas of science literacy*. Washington, DC: American Association for the Advancement of Science.

Project 2061 (American Association for the Advancement of Science). (1993). *Benchmarks for science literacy*. New York: Oxford University Press.

Reiss, M. J. (2009). The relationship between evolutionary biology and religion. *Evolution, 63*(7), 1934–1941.

Richardson, J. (2004). *Nietzsche's new Darwinism*. Oxford: Oxford University Press.

Roman Catholic Church (1996). *Magisterium is concerned with question of evolution for it involves conception of man*. Pope John Paul II. Message to Pontifical Academy of Sciences, October 22, 1996.

Runco, M. A. (1991). *Divergent thinking*. Norwood, N.J: Ablex Pub. Corp.

Russell, B. (1959). *Elaine Grand interview with Bertrand Russell. Transcript*. Montreal: Canadian Broadcasting Company.

Rutherford, F. J., & Ahlgren, A. (1990). *Science for all Americans*. New York: Oxford University Press.

Scott, E. C. (2009). Evolution vs. creationism: An introduction. Westport, CT: Greenwood Press.

Settelmaier, E. (2010). The conflict on Genesis: Building an integral bridge between creation and evolution. *Cultural Studies of Science Education, 5*(1), 243–249.

Shankar, G., & Skoog, G. (1993). Emphasis given evolution and creationism by Texas high school biology teachers. *Science Education, 77*(2), 221–233.

Sheehan, T. (1986). *The first coming: How the kingdom of God became Christianity*. New York: Random House.

Shipton, Mother, & Mee, A. (1989). *Mother Shipton's prophecies: The earliest published editions of 1641, 1684 and 1686: Together with an introduction to which is added, The story of Knaresborough: Ancient Yorkshire town of her birth*. Maidstone: George Mann.

Sinatra, G. M., & Pintrich, P. R. (2003). *Intentional conceptual change*. Mahwah: L. Erlbaum.

Sinatra, G. M., Southerland, S. A., McConaughy, F., & Demastes, J. W. (2003). Intentions and beliefs in students' understanding and acceptance of biological evolution. *Journal of Research in Science Teaching, 40*(5), 510–528.

Skoog, G. (1979). Topic of evolution in secondary school biology textbooks: 1900–1977. *Science Education, 63*(5), 621–640.

Skoog, G. (1984). The coverage of evolution in high school biology textbooks Published in the 1980s. *Science Education, 68*(2), 117–128.

Skoog, G. (2005). The coverage of human evolution in high school biology textbooks in the 20th Century and in current state science standards. *Science Education, 14*(3–5), 3–5.

Smith, M., & Siegel, H. (2004). Knowing, believing, and understanding: What goals for science education? *Science Education, 13*, 553–582.

Snow, C. P. (1962). *The two cultures and the scientific revolution; the Rede lecture, 1959.* Cambridge: University Press.

Spindler, G. D. (1987). *Education and cultural process: Anthropological approaches.* Prospect Heights: Waveland Press.

Stanford University. (2008). *Against ignorance: Science education in the 21st century: A conversation with Richard Dawkins and Lawrence Krauss.* Podcast and transcript retrieved July 20, 2009, from http://itunes.stanford.edu/

Swidler, A. (1986). Culture in action: Symbols and strategies. *American Sociological Review, 51*(April), 273–286.

Swidler, A. (2001). *Talk of love: How culture matters.* Chicago: University of Chicago Press.

Tillich, P. (1952). *The courage to be.* New Haven: Yale University Press.

Tough, P. (2008). *Whatever it takes: Geoffrey Canada's quest to change Harlem and America.* Boston: Houghton Mifflin.

Toumey, C. P. (1994). *God's own scientists: Creationists in a secular world.* New Brunswick: Rutgers University Press.

Toumey, C. P. (2004). Introduction. In S. Coleman & L. Carlin (Eds.), *The cultures of creationism: Anti-evolution in English-speaking countries.* Aldershot: Ashgate.

United States District Court for the Middle District of Pennsylvania: Case no. 04cv2688 Judge Jones; Tammy Kitzmiller, et al. plaintiffs v. Dover Area School District, et al., (2005) *defendants.* S.l: s.n..

Valencia, R. R. (2010). *Dismantling contemporary deficit thinking: Educational thought and practice.* London: Routledge.

Varenne, H. (2007). Difficult collective deliberations: Anthropological notes toward a theory of education. *Teachers College Record, 109*(7), 1559–1588.

Villegas, A. (2007). Dispositions in teacher education. *Journal of Teacher Education, 58*(5), 370–380.

Watson, J., Wilson, E. O., Rose, C., & WNET (Television station: New York). (2005). *The Charlie Rose* (Show: #11248). New York: Charlie Rose.

Weber, M. (1946). The social psychology of the world religions. In H. H. Gerth & C. Wright Mills (Eds.), *Max Weber* (pp. 267–301). New York: Oxford University Press.

White, C. J., & Ralkowski, M. (2005). *Time and death: Heidegger's analysis of finitude.* Aldershot: Ashgate Pub.

Wile, J. L., Durnell, M. F., & Sheridan Books. (2000). *Exploring creation with biology.* Anderson, Ind: Apologia Educational Ministries.

Willis, P. E. (1978). *Learning to labour: How working class kids get working class jobs.* Farnborough: Saxon House.

Wittgenstein, L. (1953). *Philosophical investigations.* New York: Macmillan.

Wrathall, M. A. (2006). *How to read Heidegger.* New York: W.W. Norton.

Wuthnow, R. (2005). *America and the challenges of religious diversity.* Princeton: Princeton University Press.

Wuthnow, R. (2009). No contradictions here: Science religion and the culture or all reasonable possibilities. In H. W. Attridge & R. L. Numbers (Eds.), *The religion and science debate: Why does it continue?* New Haven: Yale University Press.

Zimmerman, M. (1987). The evolution-creation controversy: Opinions of Ohio high school biology teachers. *The Ohio Journal of Science, 87*(4), 115–125.

Index

A
abgrund, 160
Abortion, 97
Abysmal reasoning, 47, 160
Abyss, 42, 44, 46, 152, 158, 160, 163–165
ACLU, 112
Active and authentic learning, 105
Administrative pressures, 119
Advanced placement testing, 126
Aesthetics, 3, 155, 166, 175
Agnostics, 51, 54, 55, 61, 65, 132, 155
All the theories, 16, 63, 116, 124, 127, 128,
    135, 179
Alternative assignment, 137
American Association for the Advancement
    for the Advancement of Science, 19
American Association for the Advancement of
    Science (AAAS), 24–25, 49,
    105, 167
Answers in Genesis, 5, 18, 28, 34, 57, 59,
    65, 108, 111–113, 157, 176
Anthropology, 3
Anthropology of science education, 3
Apokálypsis, 4, 45
"Approved" videos concerning evolution,
    116, 124
A Private Universe, 172
Atheists, 34, 51, 54, 55, 58, 61, 65, 66, 114,
    129, 132, 154, 155
Authentic, 46, 47, 105

B
Behe, 172
Being-in-the-world, 22, 26, 41, 47, 158, 159,
    165, 169, 177

Biblical literalism, 39, 40, 45, 57, 132
Big Bang, 43, 47, 138, 139
Biological Sciences Curriculum Study, 17
Biology classes, 26, 28, 67, 101, 103, 130, 134
Biology department, 9, 11, 98, 99, 110
Both/and, 30, 60–64, 66–68, 117, 118, 120,
    127, 131, 132, 135, 160, 161,
    168, 169, 171, 179
Boundary lines, 45
Breakdowns, 16
British Association for the Advancement of
    Science, 24
British Royal Society, 24

C
Campus Christian Center, 35, 37, 110
Campus newspaper, 98
Campus preachers, 97
Cartesian space, 150
Catholic, 47, 53, 57, 62–64, 116, 118, 124,
    126, 130, 131, 133, 134, 136, 137, 139,
    144, 146, 147, 178
Conceptual change, 19, 22, 23, 25, 30, 68
Conservative community norms, 111
Constructivism, 168
Controversy, 9–11, 20, 47, 102, 103, 110, 114,
    118, 130, 133, 135, 138, 142, 145
Craft knowledge, 13
Creationism, 5–7, 10, 14, 16, 18, 20, 21, 26,
    28, 39, 41, 43, 52, 53, 56, 61–66,
    97–100, 107, 110, 111, 123, 127, 128,
    134, 136, 137, 141, 143, 151, 152, 158,
    167, 177–179
Creationist faculty, 99
Creationist science faculty, 125

CPSIA information can be obtained
at www.ICGtesting.com
Printed in the USA
LVOW13*1429040117

519722LV00012B/245/P